T0269643

Spectral Radius of Graphs

Spectral Radius of Graphs

Dragan Stevanović

AMSTERDAM • BOSTON • HEIDELBERG • LONDON
NEW YORK • OXFORD • PARIS • SAN DIEGO
SAN FRANCISCO • SINGAPORE • SYDNEY • TOKYO

Academic Press is an imprint of Elsevier

Academic Press is an imprint of Elsevier
32 Jamestown Road, London NW1 7BY, UK
525 B Street, Suite 1800, San Diego, CA 92101-4495, USA
225 Wyman Street, Waltham, MA 02451, USA
The Boulevard, Langford Lane, Kidlington, Oxford OX5 1GB, UK

British Library Cataloguing in Publication Data
A catalogue record for this book is available from the British Library

Library of Congress Cataloging-in-Publication Data
A catalog record for this book is available from the Library of Congress

ISBN: 978-0-12-802068-5

For information on all Academic Press publications
visit our web site at store.elsevier.com

Working together
to grow libraries in
developing countries

www.elsevier.com • www.bookaid.org

To my wife Sanja and to my mentor Dragoš, both of whom, independently of each other, forced me to write this book

CONTENTS

The aim of this book is to provide an overview of important developments on the spectral radius λ_1 of adjacency matrix of simple graphs, obtained in the last 10 years or so. Most of the presented results are related to the Brualdi-Solheid problem [24], which asks to characterize graphs with extremal values of the spectral radius in a given class of graphs. As a careful reader will easily find out, this usually means characterizing graphs with the maximum spectral radius—prevailing reason being that the Rayleigh quotient, the basic building block of most proofs, allows one to check whether the spectral radius has increased after transforming a graph, but not whether it has decreased. Despite the scarcity of lemmas on the decrease of the spectral radius, increase of interest in graphs with the minimum spectral radius is motivated by the recently discovered relation [156]

$$\tau_c = \frac{1}{\lambda_1}$$

between the epidemic threshold τ_c for the effective infection rate of a SIS-type network infection and the network's spectral radius. As the network becomes virus-free in the steady state if the effective infection rate is smaller than τ_c, the task of constructing a more resistant network obviously translates to the task of constructing a network with λ_1 as small as possible, giving impetus to the minimum part of the Brualdi-Solheid problem.

The book is primarily intended for a fellow research mathematician, aiming to make new contributions to the spectral radius of graphs. The focus of presentation is not only on the overview of recent results, but also on proof techniques, conjectures, and open problems. For the impatient reader, perhaps the best starting points are the entries "conjecture" and "open problem" in the index at the end of the book. Otherwise, Chapter 2 is devoted to the study of properties of the components of the principal eigenvector corresponding to λ_1 that will be used in several occasions later in the book. Chapters 3 and 4 deal with the instances of the Brualdi-Solheid problem. Chapter 3 presents the results on the spectral radius of graphs belonging to standard graph classes, such as graphs with a given degree sequence or planar graphs. On the other hand, graph classes in Chapter 4 are mostly defined as the set of graphs having the same value of an integer-valued graph invariant, such as the diameter or the domination number.

The book is reasonably self-contained, with necessary preliminary results collected in a short, introductory chapter, but do note that it assumes some prior familiarity with graph theory (more) and linear algebra (less). It could be used in teaching, as part of a beginning graduate course or an advanced undergraduate course, including also courses within the research experience for undergraduates programs, but do also note that it lacks exercises (unless you treat conjectures and open problems as exercises).

I hope it is understandable that in a book of this size, one cannot possibly cover all the interesting new developments in the theory of graph spectra (and not even everything that has been published on the spectral radius of graphs!) in the last 10 years or so. For the reader looking forward to expand his/her knowledge of graph spectra, some further reading may be suggested. Results on the spectral properties of directed graphs are well covered in a survey paper by Brualdi [22]. Nikiforov has surveyed his results on extremal spectral graph theory [114], although some of his results are covered here as well. Results on the spectral radius of weighted graphs are more oriented toward the spectral theory of nonnegative matrices than to graph theory; the reader is, thus, referred to Chapter 6 of Friedland's manuscript on matrices [63]. If the reader is looking for a textbook covering a wider array of topics in spectral graph theory, then good choices are the books by Cvetković et al. [47], by Van Mieghem [155], or by Brouwer and Haemers [21].

At the end, I would like to acknowledge kind hospitality of the Max Planck Institute for Mathematics in Sciences in Leipzig during the final stages of writing this book. I am grateful to Türker Bıyıkoğlu, Josef Leydold, Sebastian Cioabă, and Kelly Thomas for permissions to use some of the proofs from [17, 31, 33, 64] without significant change. I am also very much grateful to my family—Sanja, Djordje, and Milica—for all their love, support, and patience while this book was being materialized from an idea to a reality.

Dragan Stevanović
Niš & Leipzig, June 2014

CHAPTER 1

Introduction

This short, introductory chapter contains definitions and tools necessary to follow the results presented in the forthcoming chapters. We will cover various graph notions and invariants, adjacency matrix, its eigenvalues and its characteristic polynomial, and some standard matrix theory tools that will be used later in proofs.

1.1 GRAPHS AND THEIR INVARIANTS

A simple graph is the pair $G = (V, E)$ consisting of the vertex set V with $n = |V|$ vertices and the edge set $E \subseteq \binom{V}{2}$ with $m = |E|$ edges. Often in the literature, n is called the order and m the size of G. Simple graphs contain neither directed nor parallel edges, so that an edge $e \in E$ may be identified with the pair $\{u, v\}$ of its distinct endvertices $u, v \in V$. We will write shortly uv for the set $\{u, v\}$. We will also use $V(G)$ and $E(G)$ to denote the vertex set and the edge set of G, if they have not been named already. To simplify notation, we will omit graph name (usually G), whenever it can be understood from the context.

For a vertex $u \in V$, the set of its neighbors in G is denoted as

$$N_u = \{v \in V : uv \in E\}.$$

The degree of u is the number of its neighbors, i.e., $\deg_u = |N_u|$. The maximum vertex degree Δ and the minimum vertex degree δ for G are defined as

$$\Delta = \max_{u \in V} \deg_u, \qquad \delta = \min_{u \in V} \deg_u.$$

Graph G is said to be d-regular graph, or just regular, if all of its vertices have degree equal to d.

A sequence $W: u = u_0, u_1, \ldots, u_k = v$ of vertices from V such that $u_i u_{i+1} \in E$, $i = 0, \ldots, k - 1$, is called a walk between u and v in G of length k. Two vertices $u, v \in V$ are connected in G if there exists a walk between them in G, and the whole graph G is connected if there exists a walk between any two of its vertices.

Spectral Radius of Graphs. http://dx.doi.org/10.1016/B978-0-12-802068-5.00001-4

The distance $d(u, v)$ between two vertices u, v of a connected graph G is the length of the shortest walk between u and v in G. The eccentricity ecc_u of a vertex $u \in V$ is the maximum distance from u to other vertices of G, i.e.,

$$ecc_u = \max_{v \in V} d(u, v).$$

The diameter D and the radius r of G are then defined as

$$D = \max_{u \in V} ecc_u, \qquad r = \min_{u \in V} ecc_u.$$

Graph $H = (V', E')$ is a subgraph of $G = (V, E)$ if $V' \subseteq V$ and $E' \subseteq E$. If $V' = V$, we say that H is the spanning subgrah of G. On the other hand, if $E' = \binom{V'}{2} \cap E$, i.e., if H contains all edges of G whose both endpoints are in H, we say that H is the induced subgraph of G. If U is a subset of vertices of $G = (V, E)$, we will use $G - U$ (or just $G - u$ if $U = \{u\}$) to denote the subgraph of G induced by $V \setminus U$. If F is a subset of edges of G, we will use $G - F$ (or just $G - e$ if $F = \{e\}$) to denote the subgraph $(V, E \setminus F)$.

A subset $C \subseteq V$ is said to be a clique in G if $uv \in E$ holds for any two distinct vertices $u, v \in C$. The clique number ω of G is the maximum cardinality of a clique in G.

A subset $S \subseteq V$ is said to be an independent set in G if $uv \notin E$ holds for any two distinct vertices $u, v \in S$. The independence number α of G is the maximum cardinality of an independent set in G.

A function $f: V \rightarrow Z$, for arbitrary set Z, is said to be a coloring of G if $f(u) \neq f(v)$ whenever $uv \in E$. The chromatic number χ is the smallest cardinality of a set Z for which there exists a coloring $f: V \rightarrow Z$. Alternatively, as $f^{-1}(z)$, $z \in Z$, is necessarily an independent set, the chromatic number χ may be equivalently defined as the smallest number of parts into which V can be partitioned such that any two adjacent vertices belong to distinct parts.

A set D of vertices of a graph G is a dominating set if every vertex of $V(G) \setminus D$ is adjacent to a vertex of S. The *domination number* γ of G is the minimum cardinality of a dominating set in G.

A set M of disjoint edges of G is a matching in G. The matching number ν of G is the maximum cardinality of a matching in G.

Given two graphs $G = (V, E)$ and $G' = (V', E')$, the function $f: V \rightarrow V'$ is an isomorphism between G and G' if f is bijection and for each $u, v \in V$

holds $\{u, v\} \in E$ if and only if $\{f(u), f(v)\} \in E'$. If there is an isomorphism between G and G', we say that G and G' are isomorphic and denote it as $G \cong G'$. In case G and G' are the one and the same graph, then we have an automorphism.

Further, a function $i \colon G \to \mathbb{R}$ is a graph invariant if $i(G) = i(G')$ holds whenever $G \cong G'$. In other words, the value of i depends on the structure of a graph, and not on the way its vertices are labeled. All the values mentioned above

$$n, m, \Delta, \delta, D, r, \omega, \alpha, \chi, \gamma, \nu$$

are examples of graph invariants. Graph theory, actually, represents a study of graph invariants and in this book the focus will be on yet another graph invariant—the spectral radius of a graph, which is defined in the next section.

We will now define several types of graphs that will appear throughout the book. The path P_n has vertices $1, \ldots, n$ and edges of the form $\{i, i + 1\}$ for $i = 1, \ldots, n - 1$. The cycle C_n is the graph obtained from P_n by adding edge $\{n, 1\}$ to it. The complete graph K_n has vertices $1, \ldots, n$ and contains all edges ij for $1 \leq i < j \leq n$. The complete bipartite graph K_{n_1, n_2} consists of two disjoint sets of vertices V_1, $|V_1| = n_1$, and V_2, $|V_2| = n_2$, and all edges $v_1 v_2$ for $v_1 \in V_1$ and $v_2 \in V_2$. The star S_n is a shortcut for the complete bipartite graph $K_{1,n-1}$. The complete multipartite graph $K_{n_1, \ldots, n_p}$ consists of disjoint sets of vertices V_i, $|V_i| = n_i$, $i = 1, \ldots, p$, and all edges $v_i v_j$, $v_i \in V_i$, $v_j \in V_j$, for $i \neq j$. The Turán graph $T_{n,p} \cong K_{\lceil n/p \rceil, \ldots, \lceil n/p \rceil, \lfloor n/p \rfloor, \ldots, \lfloor n/p \rfloor}$ is the $(p + 1)$-clique-free graph with the maximum number of edges [151]. The complete split graph $CS_{n,p} = K_{n-p,1,\ldots,1}$ consists of an independent set of $n - p$ vertices and a clique of p vertices, such that each vertex of the independent set is adjacent to each vertex of the clique.

The coalescence $G \cdot H$ of two graphs G and H with disjoint vertex sets is obtained by selecting a vertex u in G and a vertex v in H and then identifying u and v. The kite $KP_{s,r}$ is a coalescence of the complete graph K_s and the path P_r, where one endpoint of P_r is identified with an arbitrary vertex of K_s. The lollipop $CP_{s,r}$ is a coalescence of the cycle C_s and the path P_r, where one endpoint of P_r is identified with an arbitrary vertex of C_s. The bug Bug_{p,q_1,q_2} is obtained from the complete graph K_p by deleting its edge uv, and then by identifying u with an endpoint of P_{q_1} and v with an endpoint of P_{q_2}. The bag $Bag_{p,q}$ is obtained from the complete graph K_p by replacing its edge uv with a path P_q. The pineapple $PA_{n,q}$ is a graph with n vertices consisting of a

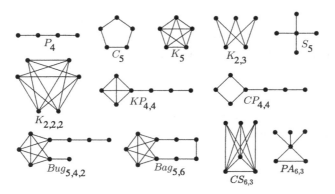

Figure 1.1 Examples of graph drawings.

clique on q vertices and an independent set on the remaining $n - q$ vertices, such that each vertex of the independent set is adjacent to the same clique vertex.

The complement of a graph $G = (V, E)$ is the graph $\overline{G} = \left(V, \binom{V}{2} \setminus E \right)$, so that each pair $\{u, v\}$, $u \neq v$, appears as an edge in exactly one of G and $\overline{G}$. Further, for two graphs $G = (V, E)$ and $G' = (V', E')$, their union $G \cup G'$ is a graph with the vertex set $V \cup V'$ and the edge set $E \cup E'$. kG is a shortcut for $\underbrace{G \cup G \cup \cdots \cup G}_{k}$. The join $G \vee G'$ of two graphs $G = (V, E)$ and $G' = (V', E')$ is a graph with the vertex set $V \cup V'$ and the edge set $E \cup E' \cup \{st : s \in V, t \in V'\}$.

Graphs are often depicted as drawings in which vertices are represented as points (actually, as circles with small diameter), and edges between them as simple curves (most often, as straight segments). Examples of such drawings are given in Fig. 1.1.

For other undefined notions, and for further study of the basics of graph theory, the reader is referred to [50], a modern "classical" textbook in graph theory.

1.2 ADJACENCY MATRIX, ITS EIGENVALUES, AND ITS CHARACTERISTIC POLYNOMIAL

The adjacency matrix of $G = (V, E)$ is the $n \times n$ matrix A indexed by V, whose (u, v)-entry is defined as

$$A_{uv} = \begin{cases} 1 & \text{if } uv \in E, \\ 0 & \text{if } uv \notin E. \end{cases}$$

Recall that a matrix is said to be reducible if it can be transformed to the form

$$A = \begin{bmatrix} A' & B \\ 0 & A'' \end{bmatrix},$$

where A' and A'' are square matrices, by simultaneous row/column permutations. Otherwise, A is said to be irreducible. It is easy to see that the adjacency matrix A is irreducible if and only if G is a connected graph.

Adjacency matrix is closely related to the numbers of walks between vertices of G. Namely,

Theorem 1.1. *The number of walks of length k, $k \geq 0$, between vertices u and v in G is equal to $(A^k)_{u,v}$.*

Proof. By induction on k. For $k = 0$, the unit matrix $A^0 = I$ has entries 1 and 0, equal to the numbers of walks of length 0, as these are the walks which consist of a single vertex only (so 1s for the diagonal entries and 0s for nondiagonal entries).

Assume now that the inductive hypothesis holds for some $k \geq 0$. Any walk of length k between u and v consists of an edge uz for some neighbor $z \in N_u$ and a walk of length $k - 1$ between z and v, so that, by the inductive hypothesis, the number of walks of length k between u and v is equal to

$$\sum_{z \in N_u} (A^{k-1})_{z,v} = \sum_{z \in V} A_{u,z}(A^{k-1})_{z,v} = (A^k)_{u,v}.$$

$\square$

The adjacency matrix A is a real, symmetric matrix, so that A is diagonalizable and has n real eigenvalues

$$\lambda_1 \geq \lambda_2 \geq \cdots \geq \lambda_n$$

and n real, linearly independent, unit eigenvectors $x_1, x_2, \ldots, x_n \in \mathbb{R}^n$ satisfying the eigenvalue equation

$$Ax_i = \lambda_i x_i, \qquad i = 1, \ldots, n.$$

The eigenvectors, in addition, can be chosen so as to form the orthonormal basis of $\mathbb{R}^n$, i.e., such that $x_i^{\mathsf{T}} x_j = 0$ for $i \neq j$.

The family of eigenvalues $\lambda_1, \ldots, \lambda_n$ is the spectrum of G. The multiplicity of an eigenvalue λ is the number of times it appears in the spectrum, i.e., it is the dimension of the subspace of eigenvectors corresponding to λ (this subspace is also called the eigenspace of λ). An eigenvalue is simple if its multiplicity is 1. The rank of a matrix is the maximum number of linearly independent columns of A. By the rank-nullity theorem [101, p. 199], the rank of A is n minus the multiplicity of eigenvalue zero.

The eigenvalues and orthonormal eigenvectors provide the spectral decomposition of A:

$$A = \sum_{i=1}^{n} \lambda_i x_i x_i^{\mathsf{T}}. \tag{1.1}$$

It is easy to see why this holds: let $B = A - \sum_{i=1}^{n} \lambda_i x_i x_i^{\mathsf{T}}$. Due to $x_i^{\mathsf{T}} x_k = 0$ for $i \neq k$, and $x_k^{\mathsf{T}} x_k = 1$, we have that for each x_k

$$Bx_k = (A - \sum_{i=1}^{n} \lambda_i x_i x_i^{\mathsf{T}}) x_k = \lambda_k x_k - \lambda_k x_k x_k^{\mathsf{T}} x_k = 0.$$

As B maps each basis vector to 0, we conclude that $B = 0$ holds.

Further, the entries of the adjacency matrix are 0s and 1s only, so that for $u \in V$,

$$(Ax_i)_u = \sum_{v \in N_u} (x_i)_v.$$

Thus, the eigenvalue equation for $i \in \{1, \ldots, n\}$ and $u \in V$ can also be written as

$$\lambda_i (x_i)_u = \sum_{v \in N_u} (x_i)_v. \tag{1.2}$$

One of important early properties of graph eigenvalues is their characterization of bipartiteness.

Theorem 1.2 ([128]). *A connected graph is bipartite if and only if $-\lambda_1$ is an eigenvalue of G, in which case the whole spectrum is symmetric with respect to 0. If G is bipartite, then the eigenvector of $-\lambda_1$ is obtained from* *its principal eigenvector by changing signs of the components in one part of the bipartition.*

See [53] for more details and further extensions of Sachs' theorem.

The characteristic polynomial of G is the characteristic polynomial of A:

$$P_G(\lambda) = \det(\lambda I - A) = \lambda^n + a_1\lambda^{n-1} + \cdots + a_n.$$

The eigenvalues of G are the roots of its characteristic polynomial $P_G(\lambda)$. The coefficients of the characteristic polynomial count, in a way, appearances of basic figures in G. An elementary figure is either an edge K_2 or a cycle C_q, $q \geq 3$, and a basic figure is any graph whose all connected components are elementary figures. Let $p(U)$ and $c(U)$ denote the number of components and the number of cycles contained in a basic figure U. If $\mathcal{U}_i$ denotes the set all basic figures contained in G having exactly i vertices, then, for $1 \leq i \leq n$,

$$a_i = \sum_{U \in \mathcal{U}_i} (-1)^{p(U)} 2^{c(U)}.$$

See, e.g., [43, p. 32] for the proof of this formula.

There are many formulas relating characteristic polynomial of a graph to those of its special subgraphs. Two most often encountered ones concern cut edges and coalescence.

Theorem 1.3 ([76]). *If uv is a cut edge of a connected graph G and G_1 and G_2 are the connected components of $G - uv$, such that u belongs to G_1 and v belongs to G_2, then*

$$P_G(\lambda) = P_{G_1}(\lambda)P_{G_2}(\lambda) - P_{G_1-u}(\lambda)P_{G_2-v}(\lambda).$$

Theorem 1.4 ([130]). *If $G \cdot H$ is the coalescence of G and H obtained by identifying a vertex u of G with a vertex v of H, then*

$$P_{G\cdot H}(\lambda) = P_G(\lambda)P_{H-v}(\lambda) + P_{G-u}(\lambda)P_H(\lambda) - \lambda P_{G-u}(\lambda)P_{H-v}(\lambda).$$

For further characteristic polynomial reduction formulas, see [47, Chapter 2].

We list here the spectra of some of the graphs defined in the previous section:

- The path P_n has eigenvalues $2\cos\frac{\pi i}{n+1}$, $i = 0, \ldots, n-1$. Its characteristic polynomial is $U_n(\lambda/2)$, where

$$U_n(x) = \sum_{k=0}^{\lfloor n/2 \rfloor} (-1)^k \binom{n-k}{k} x^{n-2k}$$

is the Chebyshev polynomial of the second kind;
- The cycle C_n has eigenvalues $2\cos\frac{2\pi i}{n}$, $i = 0, \ldots, n-1$. Its characteristic polynomial is $2T_n(\lambda/2) - 2$, where

$$T_n(x) = \frac{n}{2} \sum_{k=0}^{\lfloor n/2 \rfloor} \frac{(-1)^k}{n-k} \binom{n-k}{k} (2x)^{n-2k}$$

is the Chebyshev polynomial of the first kind;
- The complete graph K_n has a simple eigenvalue $n-1$ and eigenvalue -1 of multiplicity $n - 1$. Hence, $P_{K_n}(\lambda) = (\lambda - n + 1)(\lambda + 1)^{n-1}$;
- The complete bipartite graph K_{n_1, n_2} has two simple eigenvalues $\pm\sqrt{n_1 n_2}$ and eigenvalue 0 of multiplicity $n - 2$. Hence, $P_{K_{n_1,n_2}}(\lambda) = (\lambda^2 - n_1 n_2)\lambda^{n-2}$.

The characteristic polynomials of some other graph types, such as kites, lollipops, and bugs, can be obtained using Theorems 1.3 and 1.4. Nevertheless, their eigenvalues are not easily identifiable from their characteristic polynomials.

1.3 SOME USEFUL TOOLS FROM MATRIX THEORY

The celebrated Perron-Frobenius theorem can be applied to adjacency matrices of connected graphs.

Theorem 1.5 (The Perron-Frobenius theorem). *An irreducible, nonnegative $n \times n$ matrix A always has a real, positive eigenvalue λ_1, so that:*

1) $|\lambda_i| \leq \lambda_1$ *holds for all other (possibly complex) eigenvalues λ_i, $i = 2, \ldots, n$,*
2) λ_1 *is a simple zero of the characteristic polynomial $\det(\lambda I - A)$, and*
3) the eigenvector x_1 corresponding to λ_1 has positive components.

In addition, if A has a total of h eigenvalues whose moduli are equal to λ_1, then these eigenvalues are obtained by rotating λ_1 for multiples of angle

$2\pi/h$ *in the complex plane, i.e., these eigenvalues are equal to* $\lambda_1 e^{\frac{2\pi j}{h}}$ *for* $j = 0, \ldots, h - 1$.

For the proof of the Perron-Frobenius theorem see, e.g., [65, Chapter XIII].

Hence, the largest eigenvalue λ_1 of the adjacency matrix A of connected graph G is, at the same time, the spectral radius of A. The corresponding positive unit eigenvector x_1 is called the principal eigenvector of A.

Note that the principal eigenvector is the only positive eigenvector of A: if we would suppose that x' is another positive eigenvector of A, then we would have both $x_1^T x' = 0$ due to orthogonality, and $x_1^T x' > 0$ due to positivity of components of both eigenvectors.

Another useful result concerns the Rayleigh quotient:

$$\lambda_1 = \sup_{x \neq 0} \frac{x^T A x}{x^T x}. \tag{1.3}$$

Let the eigenvalues of A be ordered as $\lambda_1 \geq \cdots \geq \lambda_n$, and choose the orthonormal basis $x_1, \ldots, x_n$ such that x_i is the eigenvector corresponding to λ_i, $i = 1, \ldots, n$. Hence, if $x = \sum_{i=1}^n \alpha_i x_i$, then $x^T x_i = \alpha_i$, and from the spectral decomposition (1.1) follows that

$$\frac{x^T A x}{x^T x} = \frac{\sum_{i=1}^n \lambda_i x^T x_i x_i^T x}{x^T x} = \frac{\sum_{i=1}^n \lambda_i \alpha_i^2}{\sum_{i=1}^n \alpha_i^2} \leq \frac{\sum_{i=1}^n \lambda_1 \alpha_i^2}{\sum_{i=1}^n \alpha_i^2} = \lambda_1.$$

Equality is attained above for $x = x_1$, so that (1.3) holds.

As an immediate consequence of the Rayleigh quotient, we have that addition of an edge $e = uv$ to connected graph G strictly increases its spectral radius. Namely, if x is the principal eigenvector of G, then by (1.3)

$$\lambda_1(G + e) \geq \frac{x^T A x + 2 x_u x_v}{x^T x} > \frac{x^T A x}{x^T x} = \lambda_1(G). \tag{1.4}$$

Of course, this also means that deletion of an edge from a connected graph strictly decreases its spectral radius.

The Rayleigh quotient also enables the use of edge rotations and switching in order to increase the spectral radius of a graph.

Lemma 1.1 ([127]). *If for the vertices $r, s, t \in V$ of a connected graph $G = (V, E)$ holds that $rs \in E$, $rt \notin E$, and $x_s \leq x_t$, where x is the principal eigenvector of G, then*

$$\lambda_1(G - rs + rt) > \lambda_1(G).$$

Here the deletion of edge rs followed by addition of edge rt may be considered as the rotation of edge rs into rt.

Proof. We have

$$\lambda_1(G - rs + rt) \geq \frac{x^T A(G - rs + rt)x}{x^T x} = \frac{x^T A(G)x + 2x_r(x_t - x_s)}{x^T x} \geq \lambda_1(G).$$

However, the equality $\lambda_1(G - rs + rt) = \lambda_1(G)$ cannot hold. In such case, one would have that $x_s = x_t$ and that x is also the principal eigenvector of $G - rs + rt$. The eigenvalue equations at s in graphs G and $G - rs + rt$ would then give

$$\lambda_1 x_s = \sum_{u \in N_s} x_u,$$

$$\lambda_1 x_s = \lambda_1(G - rs + rt)x_s = -x_r + \sum_{u \in N_s} x_u,$$

implying that $x_r = 0$, which is a contradiction. Thus, the strict inequality holds

$$\lambda_1(G - rs + rt) > \lambda_1.$$

$\square$

The previous lemma easily extends to the simultaneous rotation of several edges (around different centers):

Lemma 1.2 ([46]). *Let s, t be two vertices of a connected graph $G = (V, E)$ and let $\{r_1, \ldots, r_t\} \subseteq N_s \setminus N_t$. Let G' be the graph obtained from G by rotating the edge sr_i to tr_i for $i = 1, \ldots, t$. If $x_s \leq x_t$, where x is the principal eigenvector of G, then $\lambda_1(G') > \lambda_1(G)$.*

Another lemma enables the increase of the spectral radius without changing the degree sequence of a graph.

Lemma 1.3 ([46]). *Let s, t, u, v be the four distinct vertices of a connected graph G and let $st, uv \in E(G)$, while $sv, tu \notin E(G)$. If $(x_s - x_u)(x_v - x_t) \geq 0$, where x is the principal eigenvector of G, then*

$$\lambda_1(G - st - uv + sv + tu) \geq \lambda_1(G)$$

with equality if and only if $x_s = x_u$ and $x_t = x_v$.

Proof. Let $G' = G - st - uv + sv + tu$. From the Rayleigh quotient we have that both

$$\lambda_1(G') \geq \frac{x^{\mathrm{T}} A(G')x}{x^{\mathrm{T}}x} = 2 \sum_{ij \in E(G')} x_i x_j \quad \text{and}$$

$$\lambda_1(G) = \frac{x^{\mathrm{T}} A(G')x}{x^{\mathrm{T}}x} = 2 \sum_{ij \in E(G)} x_i x_j.$$

Hence,

$$\lambda_1(G') - \lambda_1(G) \geq 2(x_s x_v + x_t x_u) - 2(x_s x_t + x_u x_v)$$
$$= 2(x_s - x_u)(x_v - x_t) \geq 0.$$

As for the equality above, it is easy to see that if $x_s = x_u$ and $x_t = x_v$, then x is also a positive eigenvector of G' corresponding to the eigenvalue $\lambda_1(G)$, which means that $\lambda_1(G') = \lambda_1(G)$. $\square$

The Perron-Frobenius theorem and the Rayleigh quotient also yield well-known bounds on the spectral radius of connected graphs:

$$\frac{2m}{n} \leq \lambda_1 \leq \Delta, \tag{1.5}$$

with equality in both inequalities if and only if the graph is regular. The first inequality between the average degree of a graph and λ_1 follows from the Rayleigh quotient by taking x to be the all-one vector j, while the second inequality between λ_1 and the maximum vertex degree Δ follows from the positivity of the principal eigenvector x_1: if $x_{1,u}$ is the maximum component of x_1, then from the eigenvalue equation (1.2)

$$\lambda_1 x_{1,u} = \sum_{v \in N_u} x_{1,v} \leq \sum_{v \in N_u} x_{1,u} = \deg_u x_{1,u} \leq \Delta x_{1,u}.$$

A few more well-known results that prove to be useful in the study of graph spectra follow.

An internal path of a graph G is a sequence of vertices $u_1, \ldots, u_k$ such that all u_i are distinct, except possibly $u_1 = u_k$; the vertex degrees satisfy

Figure 1.2 The graph $\tilde{D}_{n-1}$ [80].

$$\deg_{u_1} \geq 3, \quad \deg_{u_2} = \cdots = \deg_{u_{k-1}} = 2, \quad \deg_{u_k} \geq 3,$$

and $u_i u_{i+1} \in E(G)$ for $i = 1, \ldots, k-1$.

Lemma 1.4 ([80]). *Let G_{uv} be the graph obtained from G by inserting a new vertex of degree 2 on the edge uv of G. If uv is an edge on an internal path of G and $G \not\cong \tilde{D}_{n-1}$ (see Fig. 1.2), then*

$$\lambda_1(G_{uv}) < \lambda_1(G).$$

Theorem 1.6 (Interlacing theorem). *For a real, symmetric $n \times n$ matrix A with eigenvalues $\lambda_1 \geq \cdots \geq \lambda_n$ and its arbitrary principal submatrix B with eigenvalues $\mu_1 \geq \cdots \geq \mu_{n-k}$, obtained by deleting the same k rows and columns of A, holds*

$$\lambda_{i+k} \leq \mu_i \leq \lambda_i$$

for any $i = 1, \ldots, n-k$.

For the proof of the Interlacing theorem see, e.g., [155, art. 180] or [21, Section 2.5].

Theorem 1.7 (Gershgorin circle theorem). *Let $A = (a_{ij})$ be a complex $n \times n$ matrix, and let $R_i = \sum_{j \neq i} |a_{ij}|$ be the sum of nondiagonal entries in the ith row. Let $D(a_{ii}, R_i)$ be the closed circle in $\mathbb{C}$ with center in a_{ii} and radius R_i. Then each eigenvalue λ of A lies within at least one Gershgorin cicle $D(a_{ii}, R_i)$.*

Proof. Let x be an eigenvector corresponding to λ and let i be such that x_i has the maximum modulus among the components of x. From $x \neq 0$ follows that $|x_i| > 0$. Now from

$$\lambda x_i = (Ax)_i = \sum_{j=1}^{n} a_{ij} x_j,$$

we have

$$|\lambda - a_{ii}||x_i| = \left|\sum_{j \neq i} a_{ij}x_j\right| \leq \sum_{j \neq i} |a_{ij}||x_j| \leq R_i|x_i|.$$

By dividing this inequality with $|x_i|$, we see that λ belongs to the Gershgorin circle $D(a_{ii}, R_i)$. $\qquad\square$

Theorem 1.8 ([161]). *If A and B are two $n \times n$ Hermitian matrices, then for $i + j \leq n + 1$*

$$\lambda_{i+j-1}(A + B) \leq \lambda_i(A) + \lambda_j(B),$$

while for $i + j \geq n + 1$

$$\lambda_i(A) + \lambda_j(B) \leq \lambda_{i+j-n}(A + B).$$

CHAPTER 2

Properties of the Principal Eigenvector

We start our journey by surveying properties of the principal eigenvector x_1. Section 2.1 brings the author's previously unpublished proportionality lemma stating that the components of x_1 at the corresponding vertices in two isomorphic "outskirts" of the graph are proportional to each other, together with its application to determining the spectral radius of the rooted product in terms of the factor graphs. Section 2.2 exhibits the use of linear recurrence equations in determining the components of x_1 along a path and describes how it can be used to determine limits of spectral radii when the path is allowed to have the infinite length. Next, Section 2.3 contains estimates of the maximum and the minimum components of x_1, as well as those of their ratio, results that will be used in a later section on the spectral radius of nonregular graphs, but which are interesting in their own right. Section 2.4 answers which vertices or edges should be removed from a graph in order to decrease its spectral radius the most and provides lower bounds on the spectral radius of such subgraphs in terms of the spectral radius of the starting graph. Finally, Section 2.5 describes two generalizations of regular graphs: harmonic and semiharmonic graphs, obtained by assuming that the principal eigenvector is obtained from Aj and A^2j, respectively, where j is the all-one vector, and provides their characterization via generalizations of the Hoffman's identity for regular graphs.

2.1 PROPORTIONALITY LEMMA AND THE ROOTED PRODUCT

Let G be a rooted graph with the root v, and let H be an arbitrary simple graph. Let w_1 and w_2 be two arbitrary vertices of H and form the graph F from H and two copies of G by identifying w_1 with the root of the first copy of G and w_2 with the root of the second copy of G (see Fig. 2.1). Let x be the principal eigenvector of F. The following lemma gives a relation between the principal eigenvector components in two copies of G.

Lemma 2.1 (Proportionality lemma). *For any vertex u_1 in the first copy of G and the corresponding vertex u_2 in the second copy of G holds*

$$\frac{x_{u_1}}{x_{u_2}} = \frac{x_{w_1}}{x_{w_2}}. \tag{2.1}$$

Spectral Radius of Graphs. http://dx.doi.org/10.1016/B978-0-12-802068-5.00002-6

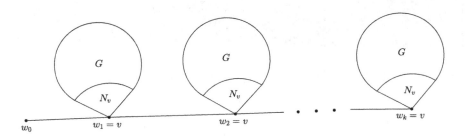

Figure 2.1 Graph F formed from H and two copies of G.

Proof. Let λ_1 be the spectral radius of F. Let A be the adjacency matrix of the vertex-deleted graph $G - v$. Since $G - v$ is an induced subgraph of F, the spectral radius of its adjacency matrix A is strictly smaller than λ_1 by (1.4). Then λ_1 is not an eigenvalue of A, so that

$$\det(\lambda_1 I - A) \neq 0.$$

As a consequence, both $[\lambda_1 I - A]$ and $[\lambda_1 I - A x|z]$ have the full rank for an arbitrary vector z, so that the system in unknown vector y

$$\lambda_1 y = Ay + z \tag{2.2}$$

has unique solution y_z by the Capelli theorem.

Let us now focus to the first copy of G and let y_1 be the $(G - v)$ part of the principal eigenvector x of F. The eigenvalue equation for λ_1, when restricted to vertices in $G - v$, becomes

$$\lambda_1 y_1 = Ay_1 + x_{w_1} b, \tag{2.3}$$

where for a vertex u of $G - v$ we have

$$b_u = \begin{cases} 1, & \text{if } u \text{ is adjacent to } v \text{ in } G, \\ 0, & \text{if } u \text{ is not adjacent to } v. \end{cases}$$

Here the vector $x_{w_1} b$ accounts in the eigenvalue equation for the principal eigenvector component of the only vertex of H that may be adjacent to a vertex of $G - v$, that is, for w_1 that was identified with v in G.

Similarly, the eigenvalue equation for λ_1, when restricted to vertices in the second copy of G, other than v, becomes

$$\lambda_1 y_2 = Ay_2 + x_{w_2} b, \tag{2.4}$$

where y_2 is the $(G - v)$ part of the principal eigenvector x.

Now, if y^* is unique solution of

$$\lambda_1 y = Ay + b,$$

then

$$y_1 = x_{w_1} y^*$$

is unique solution of (2.3), while

$$y_2 = x_{w_2} y^*$$

is unique solution of (2.4). This shows that for any vertex u_1 in the first copy of G and the corresponding vertex u_2 in the second copy of G holds

$$\frac{x_{u_1}}{x_{u_2}} = \frac{(y_1)_{u_1}}{(y_2)_{u_2}} = \frac{x_{w_1} y^*_{u_1}}{x_{w_2} y^*_{u_2}} = \frac{x_{w_1}}{x_{w_2}}.$$

$\square$

From the proof of this lemma, it is evident that

Corollary 2.1. *The ratio*

$$c = \frac{\sum_{u \in N_G(v)} x_u}{x_{w_i}}$$

does not depend on the vertex w_i of H with which v is identified:

$$c = b^T (\lambda_1 I - A_{G-v})^{-1} b.$$

Note 2.1. The previous corollary holds whenever x is an eigenvector corresponding to an eigenvalue of F that is *not* an eigenvalue of $G - v$.

Proportionality lemma may be applied, for example, to determine the spectral radius of the special case of the rooted product of graphs. The rooted product of graphs was defined by Godsil and McKay in 1978.

Definition 2.1 ([69]). Let H be a labeled graph on n vertices, and $G_1, \ldots, G_n$ be a sequence of n-rooted graphs. Then the rooted product of H by $G_1, \ldots, G_n$, denoted as $H(G_1, \ldots, G_n)$, is the graph obtained by identifying the root of G_i with the ith vertex of H for $i = 1, \ldots, n$.

Let us consider the special case of the rooted product when all G_is are isomorphic to each other as rooted graphs. In that case, we will denote $\underbrace{H(G, \ldots, G)}_{n \text{ times}}$ simply as $H(G, n)$. Let λ_1 and x be the spectral radius and the principal eigenvector of $H(G, n)$, and let $c = \sum_{u \in N_G(v)} x_u/x_w = b^T(\lambda_1 I - A_{G-v})^{-1} b$ be the constant ratio that does not depend on the vertex w of H. The eigenvalue equation for $H(G, n)$

$$\lambda_1 x = A_{H(G,n)} x$$

can be, for each vertex w of H, rewritten as

$$\lambda_1 x_w = (A_H x_H)_w + \sum_{u \in N_G(v)} x_u = (A_H x_H)_w + c x_w, \qquad (2.5)$$

where x_H is the principal eigenvector x restricted to H. Since (2.5) holds for each w, we have that actually

$$(\lambda_1 - c) x_H = A_H x_H,$$

showing that x_H is the positive eigenvector of the adjacency matrix A_H of H, and consequently, that $\lambda_1 - c$ is equal to the spectral radius μ_1 of H. Hence, the spectral radius λ_1 of $H(G, n)$ is implicitly defined by the equation

$$\lambda_1 - b^T(\lambda_1 I - A_{G-v})^{-1} b = \mu_1. \qquad (2.6)$$

This is fully in line with the results of Godsil and McKay from [69]. They showed that the characteristic polynomial of $H(G_1, \ldots, G_n)$ can be found as the determinant of a matrix, whose diagonal entries are $P_{G_i}(\lambda)$, $i = 1, \ldots, n$, while the nondiagonal (i,j) entry is $-(A_H)_{ij} P_{G_i-v}(\lambda)$. For the special case of $H(G, n)$ this reduces to

$$P_{H(G,n)}(\lambda) = P_{G-v}(\lambda)^n P_H\left(\frac{P_G(\lambda)}{P_{G-v}(\lambda)}\right),$$

a result that was first obtained by Schwenk [131]. Since the eigenvalues of $G - v$ are strictly smaller than the spectral radius of G, and hence $H(G, n)$, the spectral radius λ_1 of $H(G, n)$ then satisfies

$$\frac{P_G(\lambda_1)}{P_{G-v}(\lambda_1)} = \mu_1. \qquad (2.7)$$

In order to show that (2.7) is equivalent to (2.6), recall that the adjacency matrix A_G of G has the form

$$AG = \begin{bmatrix} A_{G-v} & b \\ b^T & 0 \end{bmatrix}$$

so that

$$P_G(\lambda) = \det(\lambda I - A_G) = \begin{vmatrix} \lambda I - A_{G-v} & -b \\ -b^T & \lambda \end{vmatrix}.$$

Since the determinant of a matrix is a linear function of any column, the right side of the above can be expressed as

$$\begin{vmatrix} \lambda I - A_{G-v} & -b \\ -b^T & \lambda \end{vmatrix} = \begin{vmatrix} \lambda I - A_{G-v} & -b \\ -b^T & 0 \end{vmatrix} + \begin{vmatrix} \lambda I - A_{G-v} & 0 \\ -b^T & \lambda \end{vmatrix}$$

$$= \det(\lambda I - A_{G-v})\det(0 - b^T(\lambda I - A_{G-v})^{-1}b)$$
$$+ \lambda \det(\lambda I - A_{G-v}),$$

where we applied the Schur identity

$$\det \begin{bmatrix} X & Y \\ Z & W \end{bmatrix} = \det X \det(W - ZX^{-1}Y)$$

to the first determinant above and the usual Laplace expansion to the second determinant. It now follows that

$$P_G(\lambda) = P_{G-v}(\lambda)\left(\lambda - \det(b^T(\lambda I - A_{G-v})^{-1}b)\right)$$
$$= P_{G-v}(\lambda)\left(\lambda - b^T(\lambda I - A_{G-v})^{-1}b\right),$$

as $b^T(\lambda I - A_{G-v})^{-1}b$ is a scalar, whose determinant is equal to its value. Therefore,

$$\frac{P_G(\lambda_1)}{P_{G-v}(\lambda_1)} = \lambda_1 - b^T(\lambda I - A_{G-v})^{-1}b$$

showing that (2.6) and (2.7) are equivalent.

Let us now briefly see two examples.

Example 2.1. Let H be a graph with n vertices and let the star $K_{1,k}$ have the center as its root. Let λ_1 and x be the spectral radius and the principal eigenvector of the rooted product $H(K_{1,k}, n)$. The principal eigenvector component of each leaf belonging to the star whose root is identified with a vertex w of H in $H(K_{1,k}, n)$ is equal to x_w/λ_1 (see Fig. 2.2).

The ratio $c = \sum_{u \in N_{K_{1,k}}(w)} x_u/x_w$ is then equal to k/λ_1, so that λ_1 and the spectral radius μ_1 of H are related by the equation

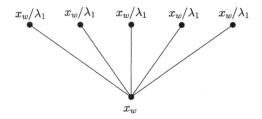

Figure 2.2 Principal eigenvector components in the star $K_{1,k}$.

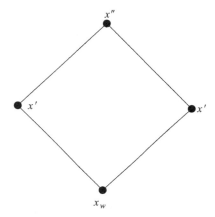

Figure 2.3 Principal eigenvector components in the cycle C_4.

$$\lambda_1 - \frac{k}{\lambda_1} = \mu_1,$$

which implies that

$$\lambda_1 = \frac{\mu_1 + \sqrt{\mu_1^2 + 4k}}{2}.$$

Example 2.2. Now, let λ_1 and x be the spectral radius and the principal eigenvector of the rooted product $H(C_4, n)$ of a graph H with n vertices and the cycle C_4 on four vertices (see Fig. 2.3).

The eigenvalue equation leads to relations

$$\lambda_1 x'' = 2x',$$
$$\lambda_1 x' = x'' + x_w$$

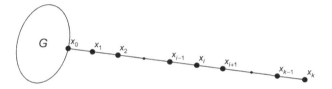

Figure 2.4 Principal eigenvector components along a pendant path.

for the vertices belonging to the cycle whose root is identified with a vertex w of H in $H(C_4, n)$. These two relations then easily imply that

$$x' = x_w \frac{\lambda_1}{\lambda_1^2 - 2},$$

$$x'' = x_w \frac{2}{\lambda_1^2 - 2},$$

so that the ratio $c = 2x'/x_w$ is equal to $\frac{2\lambda_1}{\lambda_1^2 - 2}$. Therefore, the spectral radius λ_1 of $H(C_4, n)$ is defined implicitly in terms of the spectral radius μ_1 of H by the equation

$$\lambda_1 - \frac{2\lambda_1}{\lambda_1^2 - 2} = \mu_1.$$

Although this can be solved explicitly by a computer algebra package such as Maxima (http://maxima.sourceforge.net/), the resulting formula is too complicated to be of much use.

2.2 PRINCIPAL EIGENVECTOR COMPONENTS ALONG A PATH

The eigenvalue equation allows us to use the theory of recurrence relations to determine the principal eigenvector components along a path in the graph. Such approach has been, for example, used earlier in [33, Section 2.1] and [147, Lemma 3.1].

Suppose that a pendant path P of length k has been attached to a vertex of a graph G. Let $\lambda_1 > 2$ and x be the spectral radius and the principal eigenvector of the resulting graph and denote the principal eigenvector components along the path with $x_0, \ldots, x_k$ (see Fig. 2.4). The eigenvalue equation at the ith vertex of P, $1 \le i \le k - 1$, now reads

$$\lambda_1 x_i = x_{i-1} + x_{i+1}, \tag{2.8}$$

where the value of x_0 can be taken as an initial condition. The difference between (2.8) and the usual recurrence relation is that the recurrence relation is supposed to define an infinite sequence, while P has finitely many vertices only. Hence, let us imagine that the attached path is extended to infinity with, well, invisible vertices $k + 1, k + 2, \dots$ whose associated entries $x_{k+1}, x_{k+2}, \dots$ are chosen consecutively as so to satisfy (2.8). The crucial point to notice here is that this forces the choice $x_{k+1} = 0$: on the one hand, x_{k+1} is supposed to satisfy

$$\lambda_1 x_k = x_{k-1} + x_{k+1},$$

while on the other hand, the eigenvalue equation at the last vertex of P actually reads

$$\lambda_1 x_k = x_{k-1}.$$

What do we obtain by this? Well, now the recurrence relation

$$\lambda_1 y_i = y_{i-1} + y_{i+1}, \qquad i \geq 1$$

with the initial conditions

$$\begin{aligned} y_0 &= x_0, \\ y_{k+1} &= 0 \end{aligned}$$

defines the infinite sequence $(y_i)_{i \geq 0}$ whose first $k + 1$ entries exactly correspond to the principal eigenvector components $x_0, x_1, \dots, x_k$. The solution of this linear homogeneous recurrence relation with constant coefficients (for more details on the theory of recurrence relations see, e.g., [41]) is given by

$$y_i = At^i + Bt^{-i},$$

where

$$t = \frac{\lambda_1 + \sqrt{\lambda_1^2 - 4}}{2}$$

while the constants A and B are determined from the initial conditions above:

$$\begin{aligned} A + B &= x_0, \\ At^{k+1} + Bt^{-(k+1)} &= 0, \end{aligned}$$

which imply that

$$A = -\frac{x_0}{t^{2k+2} - 1}, \qquad B = \frac{x_0\, t^{2k+2}}{t^{2k+2} - 1}.$$

Finally, for the principal eigenvector components $x_1, \ldots, x_k$ holds

$$x_i = x_0 \frac{t^{2k+2-i} - t^i}{t^{2k+2} - 1}.$$

Although we now have an explicit formula for the principal eigenvector components along a finite pendant path, the true benefit of using recurrence relations actually lies in considering an infinite pendant path attached to a vertex of G (instead of imagining it as above).

The theory of spectra of infinite graphs, with strong connections to spectra of their finite subgraphs, has been put forward in a series of papers of Mohar in the 1980s [16, 104–106]. The proper starting point for independent study is a survey by Mohar and Woess [106]. In this setting, an infinite, locally finite graph $G = (V, E)$ no longer has an adjacency matrix, but instead an adjacency operator A that acts on the Hilbert space $l^2(V)$. In the case that the vertex degrees of G are bounded by a finite number, which we assume in the rest of this section, A is a bounded (and thus, self-adjoint) operator, whose spectrum consists of

1) the point spectrum (i.e., the usual eigenvalues),
2) the continuous spectrum (the numbers λ for which $\lambda I - A$ is not injective), and
3) the residual spectrum (which is either $\emptyset$ or $\mathbb{C} \setminus \mathbb{R}$).

The value

$$r(G) = \limsup_{n \to \infty} \left[(A^n)_{u,v} \right]^{1/n}$$

does not depend on the choice of vertices u and v (for the proof see, e.g., [133, Section 6.1]), and it represents the spectral radius of the adjacency operator A. The spectral radius is part of the spectrum of A [104, Corollary 4.6], although $r(G)$ may not necessarily be an eigenvalue of A.

Definition 2.2 ([104]). A sequence of graphs $(G_n)_{n \geq 1}$ converges to G if each edge of G is contained in all but finitely many graphs from $(G_n)_{n \geq 1}$.

Connection between the spectral radius of an infinite, locally finite graph and spectral radii of its finite subgraphs is given in the following

Theorem 2.1 ([104]). *Let $(G_n)_{n \geq 1}$ be a sequence of finite subgraphs of G that converges to G. Then*

$$r(G) = \lim_{n \to \infty} \lambda_1(G_n).$$

The previous theorem allows us to determine limit of spectral radii of a sequence of finite graphs by determining the spectral radius of the infinite graph to which the sequence converges. We need one more ingredient for this, restated in more familiar terms.

Theorem 2.2 ([123]). *If G is a connected graph with adjacency operator A, then the system*

$$Ax \le \lambda x$$

for real λ and non-negative x has

1) *no solution if $\lambda < r(G)$;*
2) *infinitely many linearly independent positive solutions if $\lambda > r(G)$;*
3) *infinitely many solutions if $\lambda = r(G)$ and A is r-transient;*
4) *a unique solution x (up to positive multiples) if $\lambda = r(G)$ and A is r-recurrent and this solution satisfies $Ax = r(G)x$.*

(For the definitions of r-transient and r-recurrent operators see [133, Section 6.1]).

Hence, if we do manage to find a positive eigenvector x of A with finite norm so that

$$Ax = \lambda x,$$

then

$$\lambda \ge r(G)$$

by the point A above, and since λ is then an eigenvalue (and element of the point spectrum) of A, it has to be

$$\lambda \le r(G).$$

Thus, it has to be $\lambda = r(G)$ and then $r(G)$ is an eigenvalue of A.

So, suppose now that an infinite pendant path P_∞ has been attached to a vertex of a graph G, and let $x_0, x_1, \ldots$ be the components of the hypothetical positive eigenvector corresponding to the eigenvalue $\lambda > 2$ (see Fig. 2.5).

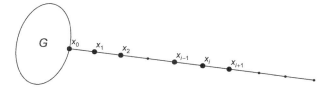

Figure 2.5 Positive eigenvector components along an infinite pendant path.

The eigenvalue equation at vertices of P_∞ gives rise to the recurrence relation

$$\lambda x_i = x_{i-1} + x_{i+1}, \qquad i \geq 1,$$

with x_0 as the initial condition. Its solution, again, is

$$x_i = At^i + Bt^{-i},$$

where

$$t = \frac{\lambda + \sqrt{\lambda^2 - 4}}{2} \qquad\qquad (2.9)$$

and suitable constants A and B. The second condition this time is the requirement that x has finite norm in $l^2(V)$, which impplies that

$$A = 0,$$

as otherwise not only the norm, but also the components x_n, $n \geq 1$, will tend to ∞, due to $t > 1$. Then, obviously, $B = x_0$ and we proved

Lemma 2.2. *If the spectral radius $\lambda > 2$ of an infinite graph G_∞ with bounded vertex degrees is its eigenvalue, then its positive eigenvector x satisfies*

$$x_i = x_0 t^{-i}$$

along any infinite pendant path of G_∞, where x_0 is the component of the vertex to which the path is attached, and t given by (2.9).

Certainly, the key to successfully finding a positive eigenvector of a graph G_∞ with infinite pendant paths lies in the structure of the graph G to which infinite pendant paths are attached. If G has a simple structure, as in the next two examples, then the spectral radius of G_∞ pops out naturally.

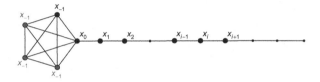

Figure 2.6 Infinite kite IK₅.

Example 2.3. Let us consider first an infinite kite IK_ω obtained by attaching an infinite pendant path to a vertex of the complete graph K_ω. Suppose that the components of the hypothetical positive eigenvector x are denoted as in Fig. 2.6, and let λ be the spectral radius of IK_ω.

The components of the vertices of K_ω, other than the one incident with the infinite path, are all denoted by x_{-1}, as similar vertices must have equal components within an eigenvector corresponding to a simple eigenvalue (see, e.g., [46, p. 44]).

Without loss of generality we may suppose that $x_0 = 1$, so that from Lemma 2.2 follows

$$x_i = t^{-i}, \qquad i \geq 0,$$

for

$$t = \frac{\lambda + \sqrt{\lambda^2 - 4}}{2}.$$

The remaining component x_{-1} then satisfies two eigenvalue equations

$$\lambda x_0 = (\omega - 1)x_{-1} + x_1,$$
$$\lambda x_{-1} = (\omega - 2)x_{-1} + x_0,$$

each of which gives an expression for x_{-1}. This yields to a further equation in terms of ω, λ, and t only:

$$\frac{\lambda - \frac{1}{t}}{\omega - 1} = \frac{1}{\lambda - \omega + 2}.$$

The key to easily solving this equation is to notice that

$$\lambda = t + \frac{1}{t},$$

which finally gives us the spectral radius of the infinite kite IK_ω:

$$\lambda = \frac{\omega - 3}{2} + \frac{\omega - 1}{2(\omega - 2)}\sqrt{\omega^2 - 4}.$$

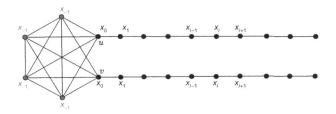

Figure 2.7 Infinite bug IB_6.

This is, due to Theorem 2.1, also the limit of the spectral radii of finite kites, a result that will be treated in more detail later in Section 4.1.

Example 2.4. Let us now consider an infinite bug IB_ω obtained by attaching two infinite pendant paths to vertices u and v of the complete graph K_ω, followed by deleting edge uv from K_ω. Suppose that the components of the hypothetical positive eigenvector x are denoted as in Fig. 2.7, and let λ be the spectral radius of IB_ω.

Vertices of K_ω, other than u and v, are mutually similar, so that they all have the same eigenvector component x_{-1}. In addition, corresponding vertices from two infinite paths are similar between each other, so that both paths have the same sequence of eigenvector components.

Again, we may freely suppose $x_0 = 1$, so that by Lemma 2.2

$$x_i = t^{-i}, \qquad i \geq 0,$$

for

$$t = \frac{\lambda + \sqrt{\lambda^2 - 4}}{2}.$$

The remaining component x_{-1} satisfies two eigenvalue equations

$$\lambda x_0 = (\omega - 2)x_{-1} + x_1,$$
$$\lambda x_{-1} = (\omega - 3)x_{-1} + 2x_0.$$

Expressing x_{-1} in two different ways from here yields an equation in terms of ω, λ, and t only

$$\frac{\lambda - \frac{1}{t}}{\omega - 2} = \frac{2}{\lambda - \omega + 3}. \tag{2.10}$$

By substituting

$$\lambda = t + \frac{1}{t}$$

in (2.10), we get a quadratic equation in t, whose solution then gives us the spectral radius of the infinite bug IB_ω:

$$\lambda = \frac{(\omega - 3)^2 + (\omega - 2)\sqrt{\omega^2 + 2\omega - 11}}{2\omega - 5}.$$

This is, due to Theorem 2.1, also the limit of the spectral radii of finite bugs, a result that will be treated in more detail later in Sections 4.6 and 4.7.

2.3 EXTREMAL COMPONENTS OF THE PRINCIPAL EIGENVECTOR

Recall that, for a graph G with adjacency matrix A, the vector Aj, where j is the all-one vector, actually represents the vector of vertex degrees of G. Hence, if we suppose that G is an r-regular graph, then $Aj = rj$, showing that r is an eigenvalue of G with the eigenvector j. The fact that j is the positive eigenvector shows, due to the Perron-Frobenius theorem, that r and j are actually the spectral radius and the principal eigenvector of any r-regular graph.

If G is not a regular graph, then the components of the principal eigenvector x necessarily differ among each other. In this section we will present bounds on $x_{\max}$ and $x_{\min}$, the maximum and the minimum components of x, and on their ratio $x_{\max}/x_{\min}$.

Papendieck and Recht [120] proved the following result for the more general case of the principal eigenvector being normalized with respect to the l_p norm, $1 \leq p < \infty$.

Theorem 2.3. *Let G be a connected graph and x its principal eigenvector, normalized so that $||x||_p = 1$. For the maximal component $x_{\max_p}$ holds*

$$x_{\max_p} \leq \left(\frac{\lambda_1^{p-2}}{1 + \lambda_1^{p-2}}\right)^{1/p},$$

with equality if and only if G is a star.

In the usual case $p = 2$, that we consider here, the previous theorem yields simply

$$x_{\max} \leq \frac{1}{2}.$$

This was improved by Cioabă and Gregory in [33, Theorem 3.2].

Theorem 2.4. *Let G be a connected graph with the principal eigenvector x. If* $\deg_i$ *denotes the degree of vertex i of G, then*

$$x_i \le \frac{1}{\sqrt{1 + \frac{\lambda_1^2}{\deg_i}}}, \tag{2.11}$$

with equality if and only if $x_i = x_{\max}$ *and G is the join of vertex i and a regular graph on* $n - 1$ *vertices.*

Proof. From the Cauchy-Schwarz inequality, we have

$$\deg_i \sum_{j \in N_i} x_j^2 = \sum_{j \in N_i} 1^2 \sum_{j \in N_i} x_j^2 \ge \left(\sum_{j \in N_i} x_j \right)^2 = (\lambda_1 x_i)^2.$$

Now

$$1 = \sum_{j \in V(G)} x_j^2 \ge x_i^2 + \sum_{j \in N_i} x_j^2 \ge x_i^2 \left(1 + \frac{\lambda_1^2}{\deg_i} \right),$$

which proves (2.11). Equality is attained if and only if $x_j = 0$ for $j \notin N_i$ and

$$x_j = \sqrt{\frac{1 - x_i^2}{\deg_i}}, \qquad j \in N_i.$$

Since G is connected, its principal eigenvector x is positive, so that the case $x_j = 0$ is impossible. Hence, all vertices different from i are adjacent to i and have the same x-component. Now, from the eigenvalue equation for $j \ne i$,

$$\lambda_1 x_j = x_i + (d_j - 1)x_j,$$

we see that all vertices different from i also have the same vertex degree, i.e., that the graph $G - i$ is regular. □

For $\deg_i = \Delta$, the inequality (2.11) yields the following upper bound

$$x_{\max} \le \frac{1}{\sqrt{1 + \frac{\lambda_1^2}{\Delta}}}.$$

The next result of Cioabă and Gregory [33] provides a lower bound on $x_{\max}$.

Theorem 2.5. *For a connected graph G holds*

$$x_{\max} \geq \frac{\lambda_1}{\sqrt{\sum_{i \in V(G)} \deg_i^2}}, \tag{2.12}$$

with equality if and only if G is regular.

Proof. For each $i \in V(G)$ we have

$$\lambda_1 x_i = \sum_{j \in N_i} x_j \leq \deg_i x_{\max}.$$

Squaring and summing for all $i \in V(G)$ we obtain

$$\lambda_1^2 \leq x_{\max}^2 \sum_{i \in V(G)} \deg_i^2,$$

which implies (2.12). Equality holds if and only if $x_j = x_{\max}$ for each $j \in V(G)$, i.e., if and only if G is regular.   □

The minimum component $x_{\min}$ can tend towards 0 in graphs with long pendant paths, as we have seen in the previous section, so that there is not much need to pursue lower bounds on $x_{\min}$. The following two results, firstly by Cioabă and Gregory [33, Theorem 3.6] and then by Nikiforov [113, Lemma 3], provide upper bounds on $x_{\min}$.

Theorem 2.6 ([33]). *If G is a graph with n vertices, m edges, and the maximum vertex degree Δ, then*

$$x_{\min} \leq \frac{\Delta - \lambda_1}{n\Delta - 2m}, \tag{2.13}$$

with equality if and only if $\deg_i = \Delta$ for each vertex i with $x_i > x_{\min}$.

Proof. Since

$$\lambda_1 \sum_{i \in V(G)} x_i = \sum_{i \in V(G)} \deg_i x_i,$$

we have that

$$(\Delta - \lambda_1) \sum_{i \in V(G)} x_i = \sum_{i \in V(G)} (\Delta - \deg_i)x_i \geq x_{\min} \sum_{i \in V(G)} (\Delta - \deg_i)$$

$$= x_{\min}(n\Delta - 2m).$$

Equality holds in the previous inequality if and only if $\deg_i = \Delta$ for each $i \in V(G)$ such that $x_i > x_{\min}$. $\square$

Note 2.2. The statement of [33, Theorem 3.6] contains a typo: the denominator contains $\sqrt{n}$ instead of n.

Theorem 2.7 ([113]). *If G is a connected graph with n vertices and the minimum vertex degree δ, then*

$$x_{\min} \leq \sqrt{\frac{\delta}{\lambda_1^2 + \delta(n - \delta)}}. \tag{2.14}$$

Proof. Let $u \in V(G)$ be a vertex of minimum degree $\deg_u = \delta$. We have

$$\lambda_1^2 x_{\min}^2 \leq \lambda_1^2 x_u^2 = \left(\sum_{v \in N_u} x_v \right)^2$$

$$\leq \delta \sum_{v \in N_u} x_v^2 \qquad \text{(by the Cauchy-Schwarz inequality)}$$

$$= \delta \left(1 - \sum_{v \in V(G) \setminus N_u} x_v^2 \right)$$

$$\leq \delta(1 - (n - \delta)x_{\min}^2),$$

which implies that

$$(\lambda_1^2 + \delta(n - \delta))x_{\min}^2 \leq \delta.$$

$\square$

Equality holds above if and only if $x_u = x_{\min}$ for each minimum degree vertex u and also $x_v = x_{\min}$ for each non-neighbor of u, while $x_v = x_{\max}$ holds for each neighbor of u. In particular, this means that the principal eigenvector x has only two distinct components, which is the case for complete bipartite graphs, for example.

Lower bounds on the principal ratio

$$\frac{x_{\max}}{x_{\min}}$$

may now be obtained by combining results from Theorems 2.5–2.7. Nevertheless, better lower bounds may be obtained starting with the Ostrowski's lower bound on the principal ratio of general nonnegative, irreducible matrices.

Theorem 2.8 ([119]). *Let $A = (a_{ij})_{n \times n}$ be a nonnegative, irreducible matrix. Let $s_i = \sum_{j=1}^{n} a_{ij}$ be the ith rowsum of A, and let $\Delta = \max_i s_i$ and $\delta = \min_i s_i$ be the maximum and the minimum rowsums of A.*

If λ_1 and x are the spectral radius and the principal eigenvector of A, then

$$\frac{x_{\max}}{x_{\min}} \geq \max\left\{\frac{\Delta}{\lambda_1}, \frac{\lambda_1}{\delta}\right\} \geq \sqrt{\frac{\Delta}{\delta}}. \qquad (2.15)$$

Proof. Let p and q be such that $s_p = \Delta$ and $s_q = \delta$. From the eigenvalue equation

$$Ax = \lambda_1 x$$

we have

$$\lambda_1 x_{\max} \geq \lambda_1 x_p = \sum_{j=1}^{n} a_{pj} x_j \geq \Delta x_{\min} \qquad (2.16)$$

and

$$\lambda_1 x_{\min} \leq \lambda_1 x_q = \sum_{j=1}^{n} a_{qj} x_j \leq \delta x_{\max}, \qquad (2.17)$$

from which the first inequality in (2.15) follows. For the second inequality it is enough to observe that from (2.16) and (2.17) follows

$$\frac{x_{\max}^2}{x_{\min}^2} \geq \frac{\Delta}{\lambda_1} \cdot \frac{\lambda_1}{\delta} = \frac{\Delta}{\delta}.$$

□

Cioabă and Gregory [33, Theorem 2.9] have improved the Ostrowski's bound.

Theorem 2.9 ([33]). *With notation as in Theorem 2.8, let* $B = \{u : s_u > \lambda_1\}$ *and* $C = \{u : s_u < \lambda_1\}$. *Let* $i \in B$ *such that* $s_i = \Delta$ *and* $j \in C$ *such that* $s_j = \delta$. *Then*

$$\frac{x_{\max}}{x_{\min}} \geq \max \left(\frac{\Delta - a_{ii} + \sum_{r \in B \setminus \{i\}} a_{ir} \frac{s_r - \lambda_1}{\lambda_1 - a_{rr}}}{\lambda_1 - a_{ii}}, \frac{\lambda_1 - a_{jj}}{\delta - a_{jj} - \sum_{r \in C \setminus \{j\}} \frac{a_{jr}(\lambda_1 - s_r)}{\lambda_1 - a_{rr}}} \right)$$

Zhang [168] has obtained another lower bound on $x_{\max}/x_{\min}$ in case of nonregular graphs.

Theorem 2.10 ([168]). *Let* G *be a connected, nonregular graph with* Δ, δ, *and* $\bar{d}$ *denoting, respectively, the maximum, the minimum, and the average vertex degree in* G. *It holds*

$$\frac{x_{\max}}{x_{\min}} \geq \frac{(\Delta - \bar{d})(\lambda_1 - \delta)}{(\Delta - \lambda_1)(\bar{d} - \delta)} \tag{2.18}$$

with equality if and only if there exists a partition of the vertex set $V(G) = V_1 \cup V_2$ *and positive integers* $0 \leq k < \Delta$ *and* $0 \leq l < \delta$ *such that each vertex in* V_1 *is adjacent to* k *vertices in* V_1 *and* $\Delta - k$ *vertices in* V_2, *while each vertex in* V_2 *is adjacent to* $\delta - l$ *vertices in* V_1 *and* l *vertices in* V_2.

Zhang has also shown that his bound (2.18) is better than the Ostrowski's bound (2.15) if and only if

$$\lambda_1 > \Delta - \frac{\sqrt{\delta}(\Delta - \delta)(\Delta - \bar{d})}{\sqrt{\delta}(\Delta - \bar{d}) + \sqrt{\Delta}(\bar{d} - \delta)}.$$

An upper bound on $x_{\max}/x_{\min}$ may be obtained by adapting the approach from the previous section about the principal eigenvector components along a path. The difference is that this time the path in question is not a pendant path, but instead an induced path between a vertex with $x_{\min}$ to a vertex with $x_{\max}$ as their principal eigenvector components.

Theorem 2.11 ([33]). *Let* G *be a connected graph, and let* $\lambda_1 > 2$ *and* x *be the spectral radius and the principal eigenvector of* G. *If* d *is the shortest distance between a vertex having* $x_{\min}$ *and a vertex having* $x_{\max}$ *as their* x *components, then*

$$\frac{x_{\max}}{x_{\min}} \leq \frac{t^{d+1} - t^{-d-1}}{t - t^{-1}}, \tag{2.19}$$

where

$$t = \frac{1}{2}\left(\lambda_1 + \sqrt{\lambda_1^2 - 4}\right).$$

Equality is attained if and only if either G is regular or there exists an induced path of length $d > 0$ whose endpoints have $x_{\min}$ and $x_{\max}$ as their x components, and the degrees of the endpoints are 1 (for $x_{\min}$) and at least 3 (for $x_{\max}$), while all other vertices on this path have degree 2.

Proof. Let u be a vertex of G with $x_{\min}$ as its x component, and v be a vertex with $x_{\max}$ as its x component, and let

$$W : u = v_0, v_1, \ldots, v_d = v$$

be the shortest path of length d in G between u and v. Note that the subgraph of G induced by vertices on W is isomorphic to a path P_d: if any two vertices v_i and v_j with $i + 2 \leq j$ would be adjacent in G, then

$$W' : u = v_0, \ldots, v_{i-1}, v_i, v_j, v_{j+1}, \ldots, v_d = v$$

would be a shorter walk between u and v than W, a contradiction.

Since we do not know in advance whether the vertices on W are adjacent to vertices not on W, the eigenvalue equation $Ax = \lambda_1 x$ this time leads only to a chain of inequalities:

$$\lambda_1 x_{v_0} \geq x_{v_1}, \tag{2.20}$$

$$\lambda_1 x_{v_k} \geq x_{v_{k-1}} + x_{v_{k+1}}, \qquad k = 1, \ldots, d-1, \tag{2.21}$$

where equality holds if and only if v_0 has degree 1, while vertices $v_1, \ldots, v_{d-1}$ have degree 2. By introducing v_{-1} with $x_{v_{-1}} = 0$, (2.20) and (2.21) can be more conveniently rewritten as

$$\begin{bmatrix} x_{v_k} \\ x_{v_{k-1}} \end{bmatrix} \leq \begin{bmatrix} \lambda_1 & -1 \\ 1 & 0 \end{bmatrix} \begin{bmatrix} x_{v_{k-1}} \\ x_{v_{k-2}} \end{bmatrix}, \qquad k = 0, \ldots, d,$$

which immediately yields

$$\begin{bmatrix} x_{v_k} \\ x_{v_{k-1}} \end{bmatrix} \leq \begin{bmatrix} \lambda_1 & -1 \\ 1 & 0 \end{bmatrix}^k \begin{bmatrix} x_{v_0} \\ x_{v_{-1}} \end{bmatrix}, \qquad k = 0, \ldots, d. \tag{2.22}$$

Since the eigenvalues of $\begin{bmatrix} \lambda_1 & -1 \\ 1 & 0 \end{bmatrix}$ are t and t^{-1}, while the corresponding eigenvectors are $\begin{bmatrix} 1 \\ t^{-1} \end{bmatrix}$ and $\begin{bmatrix} 1 \\ t \end{bmatrix}$, we have that

$$\begin{bmatrix} \lambda_1 & -1 \\ 1 & 0 \end{bmatrix} = P \begin{bmatrix} t & 0 \\ 0 & t^{-1} \end{bmatrix} P^{-1}, \qquad \text{where } P = \begin{bmatrix} 1 & 1 \\ t^{-1} & t \end{bmatrix}.$$

Equation (2.22) now becomes

$$\begin{bmatrix} x_{v_k} \\ x_{v_{k-1}} \end{bmatrix} \le P \begin{bmatrix} t^k & 0 \\ 0 & t^{-k} \end{bmatrix} P^{-1} \begin{bmatrix} x_{v_0} \\ x_{v_{-1}} \end{bmatrix}, \qquad k = 0, \ldots, d.$$

Recalling that $x_{v_{-1}} = 0$, from here we obtain

$$x_{v_k} \le \frac{t^{k-1} - t^{-k+1}}{t - t^{-1}} x_{v_0}, \qquad k = 0, \ldots, d.$$

Equation (2.19) now follows by observing that $x_{v_0} = x_{\min}$ and $x_{v_d} = x_{\max}$.

The case of equality easily follows by observing that equality holds in (2.20) and (2.21) if and only if vertex v_0 has degree 1, while vertices $v_1, \ldots, v_{d-1}$ have degree 2. The remaining conclusion $\deg_{v_d} \ge 3$ follows from

$$\lambda_1 x_{\max} = \lambda_1 x_{v_d} = \sum_{j \in N_{v_d}} x_j \le \deg_{v_d} x_{\max}$$

and the assumption $\lambda_1 > 2$. $\square$

For the end of this section, let us mention a conjecture of Cioabă and Gregory [33]. Let $pr(n)$ be the maximum value of the ratio $x_{\max}/x_{\min}$ among connected graphs with n vertices. By a computer search it has been observed that, for $3 \le n \le 9$, $pr(n)$ is always attained by one of the two kite graphs $KP_{s,r}$, obtained by identifying a vertex of the complete graph K_s with an endvertex of the path P_r, and such that $s = \lceil \frac{n+1}{4} \rceil + \epsilon$, where $\epsilon \in \{1, 2\}$. This has led Cioabă and Gregory [33] to propose

Conjecture 2.1 ([33]). *For any $n \ge 3$, the graph G with $\frac{x_{\max}}{x_{\min}} = pr(n)$ is a kite graph.*

2.4 OPTIMALLY DECREASING SPECTRAL RADIUS BY DELETING VERTICES OR EDGES

Spectral properties of matrices related to graphs have a considerable number of applications in the study of complex networks (see, e.g., [155, Chapter 7]

for further references). One such application of the spectral radius of adjacency matrix arises in the study of virus spread. In a susceptible-infectious-susceptible type of network infection, the long-term behavior of the infection in the network is determined by a phase transition at the epidemic threshold

$$\tau_c = \frac{1}{\lambda_1} :$$

if the effective infection rate is strictly smaller than τ_c, then the virus eventually dies out, while if it is strictly larger than τ_c then the network remains infected [156]. A question that naturally arises and that was studied in [157] is how to mostly increase network's epidemic threshold τ_c, i.e., how to mostly decrease graph's spectral radius λ_1 by removing a fixed number of its vertices or edges.

As with majority of interesting graph problems, these two problems—removing vertices or removing edges from a graph to mostly decrease its spectral radius—also happen to be NP complete, as shown in [157].

Problem 2.1 (NSRM problem). Given a graph $G = (V, E)$ and an integer $p < |V|$, determine which subset V' of p vertices needs to be removed from G, such that the spectral radius of $G - V'$ has the smallest spectral radius among all possible subgraphs that can be obtained by removing p vertices from G.

Problem 2.2 (LSRM problem). Given a graph $G = (V, E)$ and an integer $q < |E|$, determine which subset E' of q edges needs to be removed from G, such that the spectral radius of $G - E'$ has the smallest spectral radius among all possible subgraphs that can be obtained by removing q edges from G.

Theorem 2.12 ([157]). *The NSRM problem is NP-complete.*

Proof. We will prove this theorem by polynomially reducing a known NP-complete problem to the NSRM problem. The NP-complete problem that we will rely on is the *independent set problem* [67]: given a graph $G = (V, E)$ and a positive integer $k \leq |V|$, is there an independent set V' of vertices in G such that $|V'| \geq k$?

Note that the smallest possible spectral radius of a graph equals 0, which is obtained for and only for a graph without any edges. Hence, to solve the independent set problem it suffices to solve the NSRM problem with $p = |V| - k$, such that the spectral radius of the resulting vertex-deleted subgraph $G - V'$ is smallest possible: if $\lambda_1(G - V') = 0$, then $V \setminus V'$ is an independent set of k vertices in G; if $\lambda_1(G - V') > 0$, then no independent set with at least k vertices exists in G. $\qquad \square$

Before we prove that the LSRM problem is also NP complete, we need the following auxiliary lemma.

Lemma 2.3 ([157]). *The path P_n has the smallest spectral radius among all graphs with n vertices and $n - 1$ edges.*

Proof. It is long known that P_n has the smallest spectral radius among trees and, more generally, connected graphs on n vertices (see, e.g., [43, p. 21] or [155, p. 125]).

Suppose, therefore, that G is a disconnected graph with n vertices and $n - 1$ edges, and let $G_1, \ldots, G_k$, $k \geq 2$, be its connected components. If each G_i, $i = 1, \ldots, k$, is a tree, then

$$|E(G_i)| = |V(G_i)| - 1,$$

and by summing these equalities for all $i = 1, \ldots, k$, we obtain

$$n - 1 = |E(G)| = \sum_{i=1}^{k} |E(G_i)| = \left(\sum_{i=1}^{k} |V(G_i)| \right) - k = n - k,$$

which is a contradiction with $k \geq 2$.

Hence, at least one of $G_1, \ldots, G_k$ contains a cycle C as its subgraph. By the monotonicity of spectral radius we then have

$$\lambda_1(G) \geq \lambda_1(C) = 2 > 2 \cos \frac{\pi}{n + 1} = \lambda_1(P_n).$$

$\qquad \square$

Theorem 2.13 ([157]). *The LSRM problem is NP complete.*

Proof. In this case we will rely on the *Hamiltonian path problem*, another well-known NP-complete problem [67]: given a graph $G = (V, E)$, does it contain a Hamiltonian path that visits every vertex exactly once?

It then suffices to solve the LSRM problem with $q = |E| - |V| + 1$ in order to solve the Hamiltonian path problem: if for the resulting graph $G - E'$ with $|V| - 1$ edges we obtain $\lambda_1(G - E') = 2\cos\frac{\pi}{n+1}$, then $G - E'$ is a Hamiltonian path in G; if $\lambda_1(G - E') > 2\cos\frac{\pi}{n+1}$, then G does not contain a Hamiltonian path. $\qquad\qquad\qquad\qquad\qquad\qquad\qquad\qquad\qquad\qquad\qquad\qquad$ □

Due to the current absence of efficient algorithms to solve NP-complete problems (see, e.g., http://www.claymath.org/millenium-problems/p-vs-np-problem for more information on the P vs NP problem), the usual way to deal with such problems, especially in the cases of large instances, is to provide a heuristic method for finding a solution that is, hopefully, close to the optimal one. A popular choice among heuristic methods is the greedy approach which assumes that the solution is built in pieces, where at each step the locally optimal piece is selected and added to the solution. There is not necessarily a guarantee that the solution built this way will be globally optimal (unless your problem has a matroid structure—see, e.g., [39, Chapter 16]), but greedy algorithms do often find good approximations to the optimal solution.

When applied to the NSRM and LSRM problems, the greedy approach boils down to two subproblems.

Problem 2.3. Given a graph $G = (V, E)$, determine which vertex u needs to be removed from G, such that

$$\lambda_1(G - u) = \min_{v \in V(G)} \lambda_1(G - v).$$

Problem 2.4. Given a graph $G = (V, E)$, determine which edge uv needs to be removed from G, such that

$$\lambda_1(G - uv) = \min_{e \in E(G)} \lambda_1(G - e).$$

The solution to the NSRM or LSRM problem is then built in steps, where at each step we solve one of the Problems 2.3 and 2.4.

Unsurprisingly, the key to solving these two problems lies in the principal eigenvector x of G. We will show that, under suitable assumptions, spectral radius is mostly decreased by removing a vertex with the largest principal eigenvector component (for Problem 2.3) or by removing an edge with the largest product of principal eigenvector components of its endpoints (for Problem 2.4). The following argument using the numbers of closed walks, which extends to the next two subsections, is taken from [157].

Let A be adjacency matrix of a connected graph G, and let $\lambda_1 > \lambda_2 \geq \cdots \geq \lambda_n$ be the eigenvalues of A, with $x_1, x_2, \ldots, x_n$ the corresponding eigenvectors, which form the orthonormal basis. From the spectral decomposition

$$A = \sum_{i=1}^{n} \lambda_i x_i x_i^T,$$

using $x_i^T x_j = 0$ for $i \neq j$ and $x_i^T x_i = 1$ for any i, we have that

$$A^k = \sum_{i=1}^{n} \lambda_i^k x_i x_i^T.$$

When $k \to \infty$, the most important term in the above sum is $\lambda_1^k x_1 x_1^T$, provided that G is nonbipartite. In such case, we have $\lambda_1 > |\lambda_i|$ for $i = 2, \ldots, n$, and so, for any two vertices u, v of G,

$$\lim_{k \to \infty} \frac{(A^k)_{u,v}}{\lambda_1^k x_{1,u} x_{1,v}} = \lim_{k \to \infty} \frac{\sum_{i=1}^{n} \lambda_i^k x_{i,u} x_{i,v}}{\lambda_1^k x_{1,u} x_{1,v}} = 1 + \sum_{i=2}^{n} \frac{\lambda_i^k x_{i,u} x_{i,v}}{\lambda_1^k x_{1,u} x_{1,v}} = 1. \quad (2.23)$$

Remark 2.1. In case G is bipartite, let (U, V) be the bipartition of vertices of G. Then $\lambda_n = -\lambda_1$, $x_{n,u} = x_{1,u}$ for $u \in U$ and $x_{n,v} = -x_{1,v}$ for $v \in V$. Both λ_1 and λ_n are simple eigenvalues, so that $\lambda_1 > |\lambda_i|$ for $i = 2, \ldots, n-1$. Similarly as above we get

$$\lim_{k \to \infty} \frac{(A^k)_{u,v}}{\lambda_1^k x_{1,u} x_{1,v}} = 1 + \lim_{k \to \infty} (-1)^k \frac{x_{n,u} x_{n,v}}{x_{1,u} x_{1,v}}.$$

Obviously, the limit above exists only if we restrict k to range over odd or even numbers only, in which case the limit is either 0 or 2, depending on whether u and v belong to the same or different parts of the bipartition. This suggests that the same strategy will extend to bipartite graphs as well, except that the explanation will have to take into account the nonexistence of odd closed walks.

In view of (2.23), we will deliberately resort to the following *approximation*:

$$(A^k)_{u,v} \approx \lambda_1^k x_{1,u} x_{1,v} \qquad \text{for large } k.$$

Under such approximation, the total number of closed walks of large length k in G is then

$$\sum_{u \in V(G)} (A^k)_{u,u} = \sum_{u \in V(G)} \lambda_1^k x_{1,u} x_{1,u} = \lambda_1^k \sum_{u \in V(G)} (x_{1,u})^2 = \lambda_1^k.$$

2.4.1 Vertex Removal

In order to find out which vertex removal mostly decreases spectral radius, we will consider the *equivalent* question: the removal of which vertex u mostly reduces the number of closed walks in G for some large length k, under the above assumption that the number of closed walks of length k which start at vertex u is equal to $\lambda_1^k x_{1,u}^2$. Note that when we delete vertex u from G, then, besides closed walks which start at u, we also destroy closed walks which start at another vertex, but contain u as well. In addition, any closed walk that contains u may contain several occurences of u.

So, for fixed u, k, and v, let W_t denote the number of closed walks of length k which start at v and which contain u *at least* t times, $t \geq 1$. Suppose that in such a walk, vertex u appears after l_1 steps, after $l_1 + l_2$ steps, after $l_1 + l_2 + l_3$ steps, and so on, the last appearance accounted for being after $l_1 + \cdots + l_t$ steps. Here $l_1, \ldots, l_t \geq 1$. Moreover, u must appear for the last time after at most $k - 1$ steps (after k steps we are back at v), thus we may also introduce $l_{t+1} = k - (l_1 + \cdots + l_t)$ and ask that $l_{t+1} \geq 1$. Then we have

$$W_t = \sum_{l_1,\ldots,l_{t+1}} (A^{l_1})_{v,u}(A^{l_2})_{u,u} \cdots (A^{l_t})_{u,u}(A^{l_{t+1}})_{u,v}$$

$$= \sum_{l_1,\ldots,l_{t+1}} \lambda_1^k x_{1,v}^2 x_{1,u}^{2t}$$

$$= \lambda_1^k x_{1,v}^2 x_{1,u}^{2t} |\{(l_1,\ldots,l_{t+1}): l_1 \geq 1,\ldots,l_{t+1} \geq 1, l_1 + \cdots + l_{t+1} = k\}|.$$

Introducing $l_1' = l_1 - 1, \ldots, l_{t+1}' = l_{t+1} - 1$, the cardinality of the last set is equal to the number of nonnegative solutions to

$$l_1' + l_2' + \cdots + l_t' + l_{t+1}' = k - t - 1,$$

which is, in turn, equal to $\binom{(k-1-t)+t}{t} = \binom{k-1}{t}$. Therefore,

$$W_t = \binom{k-1}{t} \lambda_1^k x_{1,v}^2 x_{1,u}^{2t}.$$

Consider now a closed walk of length k starting at v which contains u *exactly j* times. Such walk is counted j times in W_1, $\binom{j}{2}$ times in W_2, $\binom{j}{3}$ times in $W_3, \ldots$, $\binom{j}{j}$ times in W_j, and using the well-known equality

$$1 = \sum_{t \geq 1} (-1)^{t-1} \binom{j}{t},$$

we see that this closed walk is counted exactly once in the expression

$$W^v = W_1 - W_2 + W_3 - \cdots + (-1)^{t-1} W_t + \cdots$$

Thus, W^v represents the number of closed walks of length k starting at v which will be affected by deleting u. From the above expression for W_t, we have

$$
\begin{aligned}
W^v &= \sum_{t \geq 1} (-1)^{t-1} \binom{k-1}{t} \lambda_1^k x_{1,v}^2 x_{1,u}^{2t} \\
&= -\lambda_1^k x_{1,v}^2 \sum_{t \geq 1} \binom{k-1}{t} (-x_{1,u}^2)^t \\
&= \lambda_1^k x_{1,v}^2 \left[1 - (1 - x_{1,u}^2)^{k-1} \right].
\end{aligned}
$$

Finally, the total number of closed walks of length k destroyed by deleting u is equal to

$$
\begin{aligned}
W &= \lambda_1^k x_{1,u}^2 + \sum_{v \neq u} W^v \\
&= \lambda_1^k x_{1,u}^2 + \lambda_1^k \sum_{v \neq u} x_{1,v}^2 \left[1 - (1 - x_{1,u}^2)^{k-1} \right] \\
&= \lambda_1^k \left[x_{1,u}^2 + (1 - x_{1,u}^2) \left[1 - (1 - x_{1,u}^2)^{k-1} \right] \right] \\
&= \lambda_1^k \left[1 - (1 - x_{1,u}^2)^k \right].
\end{aligned}
$$

The function W is increasing in $x_{1,u}$ in the interval $[0, 1]$, and we may conclude that most closed walks are destroyed when we remove the vertex with the largest principal eigenvector component. Hence, the spectral radius of G is decreased the most in such a case as well.

Alternative argument for deleting the vertex with the largest principal eigenvector component may be found in the corollary of the following theorem.

Theorem 2.14. *Let* $G = (V, E)$ *be a connected graph with* $\lambda_1(G)$ *and* x *as the spectral radius and the principal eigenvector of its adjacency matrix* $A = (a_{uv})$. *Further, let* S *be any subset of vertices of* G *and let* $\lambda_1(G - S)$ *be the spectral radius of the graph* $G - S$. *Then*

$$\frac{\left(1 - 2\sum_{s \in S} x_s^2\right) \lambda_1(G) + 2 \sum_{s \in S} \sum_{t \in S} a_{st} x_s x_t}{1 - \sum_{s \in S} x_s^2} \leq \lambda_1(G - S) < \lambda_1(G).$$

$$(2.24)$$

A variant of this theorem, although without $1 - \sum_{u \in V} x_u^2$ in the denominator, appears in [90], while its variant for a single vertex deletion is implicitly contained in the proof of [113, Lemma 4]. The proof given here is a polished version of the union of these proofs.

Proof. The second inequality above holds because of the monotonicity of the spectral radius with respect to edge addition (1.4).

We will use the Rayleigh quotient twice to prove the first inequality. Firstly, since the principal eigenvector x has unit norm, from the Rayleigh quotient we have

$$\lambda_1(G) = \frac{x^T A x}{x^T x} = 2 \sum_{uv \in E} x_u x_v.$$

Dividing the sum above into the parts corresponding to the edges within $G - S$ and the edges incident with a vertex of S, we obtain

$$\lambda_1(G) = 2 \sum_{uv \in E(G-S)} x_u x_v + 2 \sum_{s \in S} x_s \sum_{w \in N_G(s)} x_w - 2 \sum_{s \in S} \sum_{t \in S} a_{st} x_s x_t.$$

The third term in the previous equation corrects for the edges st, $s, t \in S$, that are counted twice in the second term. From the eigenvalue equation

$$\sum_{w \in N_G(s)} x_w = \lambda_1(G) x_s,$$

we further obtain

$$\left(1 - 2\sum_{s \in S} x_s^2\right) \lambda_1(G) + 2 \sum_{s \in S} \sum_{t \in S} a_{st} x_s x_t = 2 \sum_{uv \in E(G-S)} x_u x_v. \quad (2.25)$$

Now we can apply the Rayleigh quotient for the second time to the restriction $x_{V \setminus S}$ of x to $V \setminus S$ and the restriction $A_{V \setminus S}$ of A to indices in $V \setminus S$:

$$\frac{2\sum_{uv\in E(G-S)}x_u x_v}{1-\sum_{s\in S}x_s^2} = \frac{x_{V\backslash S}^T A_{V\backslash S} x_{V\backslash S}}{x_{V\backslash S}^T x_{V\backslash S}} \le \lambda_1(G-S).$$

Together with (2.25) this yields (2.24).   ∎

If we delete a single vertex s from G, i.e., $S = \{s\}$, then the term $\sum_{s\in S}\sum_{t\in S}a_{st}x_s x_t$ disappears, due to $a_{ss}=0$, and we get

Corollary 2.2. *Let $G = (V,E)$ be a connected graph with $\lambda_1(G)$ and x as its spectral radius and the principal eigenvector. If s is any vertex of G and $\lambda_1(G-s)$ is the spectral radius of the graph $G-s$, then*

$$\frac{1-2x_s^2}{1-x_s^2}\lambda_1(G) \le \lambda_1(G-s) < \lambda_1(G). \tag{2.26}$$

We can now see that if we delete the vertex s with the largest principal eigenvector component from G, then $\lambda_1(G-s)$ gets the largest "window of opportunity" to place itself within. This does not mean that $\lambda_1(G-s)$ will necessarily be close to the lower bound in (2.26), but it is certainly a better choice than the vertices for which the lower bound in (2.26) is much closer to $\lambda_1(G)$.

Another corollary may be obtained by observing that the right-hand side of (2.25) is nonnegative.

Corollary 2.3 ([90]). *Let $G = (V,E)$ be a connected graph with $\lambda_1(G)$ and x as the spectral radius and the principal eigenvector of its adjacency matrix $A = (a_{uv})$. If S is any subset of vertices of G, then*

$$\sum_{s\in S}x_s^2 \le \frac{1}{2} + \frac{1}{\lambda_1}\sum_{s\in S}\sum_{t\in S}a_{st}x_s x_t. \tag{2.27}$$

Proof. Due to the positivity of the principal eigenvector, we have in (2.25) that

$$2\sum_{uv\in E(G-S)}x_u x_v \ge 0,$$

which implies that also

$$\left(1-2\sum_{s\in S}x_s^2\right)\lambda_1(G) + 2\sum_{s\in S}\sum_{t\in S}a_{st}x_s x_t \ge 0.$$

Rearranging the terms now yields (2.27).   ∎

Cioabă [32] has recently shown that if S is an independent set of a connected graph and x is the principal eigenvector of G, then

$$\sum_{i \in S} x_i^2 \leq \frac{1}{2}.$$

This follows directly from Corollary 2.3 by noting that $a_{st} = 0$ for each $s, t \in S$.

2.4.2 Edge Removal

As in the previous subsection, we want to find out the deletion of which edge uv mostly reduces the number of closed walks in G of some large length k?

For fixed u, v, and k, let W_t denote the number of closed walks of length k which start at some vertex w and contain the edge uv at *least* t times, $t \geq 1$. Suppose that in such walk, the edge uv appears at positions $1 \leq l_1 \leq l_2 \leq \cdots \leq l_t \leq k$ in the sequence of edges in the walk, and let $u_{i,0}$ and $u_{i,1}$ be the first and the second vertex of the ith appearance of uv in the walk. Obviously, either $(u_{i,0}, u_{i,1}) = (u, v)$ or $(u_{i,0}, u_{i,1}) = (v, u)$. Then

$$
\begin{aligned}
W_t &= \sum_{w \in V} \sum_{l_1 \leq \cdots \leq l_t} (A^{l_1-1})_{w,u_{1,0}} \left(\prod_{i=2}^{t} (A^{l_i-l_{i-1}-1})_{u_{i-1,1},u_{i,0}} \right) (A^{k-l_t-1})_{u_{t,1},w} \\
&= \sum_{w \in V} \sum_{l_1 \leq \cdots \leq l_t} \lambda_1^{l_1-1} x_w x_{u_{1,0}} \left(\prod_{i=2}^{t} \lambda_1^{l_i-l_{i-1}-1} x_{u_{i-1,1}} x_{u_{i,0}} \right) \lambda_1^{k-l_t-1} x_{u_{t,1}} x_w \\
&= \sum_{w \in V} x_w^2 \sum_{l_1 \leq \cdots \leq l_t} \lambda_1^{k-t} \prod_{i=1}^{t} (x_{u_{i,0}} x_{u_{i,1}})^2 \\
&= \binom{k}{t} \lambda_1^{k-t} (2 x_u x_v)^t.
\end{aligned}
$$

The term 2 appears in front of $x_u x_v$ in the last equation as there are two ways to choose $(x_{u_{i,0}}, x_{u_{i,1}})$ for each $i = 1, \ldots, t$.

Now, the number of walks affected by deleting the link uv is equal to

$$
\begin{aligned}
W_{uv} &= \sum_{t \geq 1} (-1)^{t-1} W_t \\
&= \sum_{t \geq 1} (-1)^{t-1} \binom{k}{t} \lambda_1^{k-t} (2 x_u x_v)^t
\end{aligned}
$$

$$= \lambda_1^k - \sum_{t \geq 0} (-1)^t \binom{k}{t} \lambda_1^{k-t} (2x_u x_v)^t$$

$$= \lambda_1^k - (\lambda_1 - 2x_u x_v)^k.$$

The function W_{uv} is increasing in $x_u x_v$ in the interval $[0, \lambda_1/2]$, and so most closed walks are destroyed when we remove the edge with the largest product of principal eigenvector components of its endpoints. Thus, the spectral radius is decreased mostly in such case as well.

The two principal eigenvector heuristics for solving Problems 2.3 and 2.4 have been extensively tested in [157]. Based on test results, it has been conjectured there that the difference in the spectral radius after optimally deleting q edges from $G = (V, E)$ is proportional to q. Another expectation from [157] is that the optimal way to delete a subset E' of q edges is to make the resulting edge-deleted subgraph $G - E'$ as regular as possible: $\lambda_1(G - E')$ is, for each such E', bounded from below by the constant average degree $\frac{2(|E|-q)}{|V|}$ of $G - E'$, and the spectral radius of nearly regular graphs is close to their average degree.

Let us conclude this section with a related open problem that appears not to have been studied in the literature so far.

Problem 2.5. What is the minimum spectral radius among connected graphs with n vertices and m edges, for given n and m?

Solution is easy in the cases of trees and unicyclic graphs: if $m = n - 1$, the minimum spectral radius $2\cos\frac{\pi}{n+1}$ is obtained for the path P_n, and if $m = n$, the minimum spectral radius 2 is obtained for the cycle C_n. Note that the point of the problem is not to provide solutions for the next obvious choices $m = n + 1$ and $m = n + 2$, for example, but to solve it in the general case when m is any fixed number between $n + 1$ and $\binom{n}{2}$.

The corresponding problem on the *maximum* spectral radius of connected graphs with n vertices and m edges is well studied. It was initially posed for possibly disconnected graphs by Brualdi and Hoffman in 1976 [14, p. 438]. They later showed that if $m = \binom{d}{2}$ for $d > 1$, then the graph with the maximum spectral radius consists of the complete graph K_d and a number of isolated vertices and conjectured that if $\binom{d}{2} < m < \binom{d+1}{2}$, the graph with the maximum spectral radius consists of the complete graph K_d to which a new vertex of degree $m - \binom{d}{2}$ is added, together with the necessary number of isolated vertices [23]. This conjecture was proved by Rowlinson [126].

However, if we restrict ourselves to connected graphs with n vertices and m edges, then the problem is still largely open. Rowlinson's proof [126] of the Brualdi-Hoffman conjecture obviously resolves the cases with $m > \binom{n-1}{2}$. For general values of m, Brualdi and Solheid [25] have proved that the connected graph with the maximum spectral radius must have a stepwise adjacency matrix, meaning that the set of vertices can be ordered in such a way that whenever $a_{ij} = 1$ with $i < j$, then $a_{hk} = 1$ for $k \leq j$, $h \leq i$ and $h < k$. Recall that a threshold graph is constructed from a single vertex by consecutively adding new vertices, such that each new vertex is adjacent to either all or none of the previous vertices. It is easy to see that a connected graph with a stepwise adjacency matrix is a threshold graph without isolated vertices (i.e., the last added vertex is adjacent to all previous vertices). Let us use the notation for such graphs from [117]: start with $G_{p_1} = K_{p_1}$ and then define recursively for $k \geq 2$

$$G_{p_1,\dots,p_{k-1},p_k} = \overline{G_{p_1,\dots,p_{k-1}}} \vee K_{p_k}.$$

The case $m = n - 1$ have been solved first by Collatz and Sinogowitz [38], and later by Lovász and Pelikán [98], who showed that the star $S_n = G_{n-1,1}$ has the maximum spectral radius among trees. Brualdi and Solheid [25] have solved the cases $m = n$ $(G_{2,n-3,1})$, $m = n + 1$ $(G_{2,1,n-4,1})$, $m = n + 2$ $(G_{3,n-4,1})$, and for all sufficiently large n, also the cases $m = n + 3$ $(G_{4,1,n-6,1})$, $m = n + 4$ $(G_{5,1,n-7,1})$ and $m = n + 5$ $(G_{6,1,n-8,1})$. Cvetković and Rowlinson [45] have further proved that for fixed $k \geq 6$, the graph with the maximum spectral radius and $m = n + k$ is $G_{k+1,1,n-k-3,1}$ for all sufficiently large n. Bell [11] has solved the case $m = n + \binom{d-1}{2} - 1$, for any natural number $d \geq 5$, by showing that the maximum graph is either $G_{d-1,n-d,1}$ or $G_{\binom{d-1}{2},1,n-\binom{d-1}{2}-2,1}$, depending on a relation between n and d. Olesky et al. [117] have extended Bell's result to $m = n + \binom{d-1}{2} - 2$ for $2n \leq m < \binom{n}{2} - 1$, and the maximum graph in this case is $G_{2,d-2,n-d-1,1}$. In the remaining cases $m = n + \binom{d-1}{2} + t - 1$, for some d and $0 < t < d - 2$, both Bell [11] and Olesky et al. [117] expect that the maximum graph is either $G_{d-t-1,1,1,t,n-d-1,1}$ or $G_{\binom{d-1}{2}+t,1,n-\binom{d-1}{2}-t-2,1}$.

Recently, Bhattacharya et al. [15] studied the problem of the maximum spectral radius among connected bipartite graphs with given number m of edges and numbers p, q of vertices in each part of the bipartition, but excluding complete bipartite graphs. They have conjectured that the maximum graph is obtained from a complete bipartite graph by adding a new vertex and a corresponding number of edges. This conjecture has been proved in [15] in the case $m \equiv -1 \pmod{r}$ for some $r \geq 2$, such that $l = \lfloor m/r \rfloor \geq r$, $p \in [r, l+1]$, and $q \in [l+1, l+1+\frac{l}{r-1}]$, in which case the maximum spectral radius is attained by the graph $K_{r,l+1} - e$ for any edge e. In general, the candidate graphs for the maximum spectral radius among connected bipartite graphs are the difference graphs [99]: for a given set of positive integers $D = \{d_1 \geq \cdots \geq d_p\}$, vertices can be partitioned as $U = \{u_1, \ldots, u_p\}$ and $V = \{v_1, \ldots, v_q\}$, such that the neighbors of u_i are $v_1, \ldots, v_{d_i}$. Some spectral properties of the candidate graphs have been studied in [2, 15].

2.5 REGULAR, HARMONIC, AND SEMIHARMONIC GRAPHS

The spectral radius and the principal eigenvector of a regular graph G are immediately known: the fact that each vertex has degree r is equivalent to the assertion

$$Aj = rj, \tag{2.28}$$

where A is the adjacency matrix of G and j is the all-one vector, showing that r is the spectral radius of G and $j/\sqrt{n}$ its principal eigenvector. In this section we will showcase two further classes of graphs for which the principal eigenvector is immediately available.

A graph G is called λ-harmonic [51] (or, shortly, harmonic) if there exists a constant λ such that

$$\sum_{v \in N_u} \deg_v = \lambda \deg_u$$

holds for each $u \in V(G)$. Since the the vector of vertex degrees of G is equal to Aj, this is equivalent to the assertion

$$A^2 j = \lambda A j, \tag{2.29}$$

showing that Aj is the eigenvector corresponding to the spectral radius λ of a harmonic graph. Further, as (2.29) is obtained from (2.28) by multiplying it with A from the left, we see that every r-regular graph is also r-harmonic.

Next, let

$$\deg_{2,u} = \sum_{v \in N_u} \deg_v$$

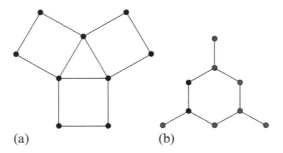

Figure 2.8 (a) A 3-harmonic graph and (b) 5-semiharmonic graph.

denote the number of walks of length two starting from u. A graph G is called λ-semiharmonic [51] (or, shortly, semiharmonic) if there exists a constant λ such that

$$\sum_{v \in N_u} \deg_{2,u} = \lambda \deg_u$$

holds for each $u \in V(G)$. The fact that G is semiharmonic is equivalent to the assertion that

$$A^3 j = \lambda A j. \tag{2.30}$$

Hence, λ and Aj are this time the spectral radius and the principal eigenvector of A^2. The spectral radius of G is then equal to $\sqrt{\lambda}$, while its corresponding eigenvector is equal to $A^2 j + \sqrt{\lambda} A j$ (check!).

Examples of a harmonic and a semiharmonic graph are depicted in Fig. 2.8.

Hoffman [78] provided another spectral characterization of regular graphs in terms of matrix polynomials. Let

$$\mu_1 > \mu_2 > \cdots > \mu_p$$

be distinct eigenvalues of the adjacency matrix A of graph G. Hoffman showed that G is a connected, regular graph if and only if

$$n \prod_{i=2}^{p} (A - \mu_i I) = \prod_{i=2}^{p} (\mu_1 - \mu_i) J \tag{2.31}$$

holds, where J is the all-one $n \times n$ matrix and I is the unit matrix. Following [52], we will present a general approach that will not only yield the proof of (2.31), but also provide corresponding characterizations for both harmonic and semiharmonic graphs.

Let $G = (V, E)$ be a finite, simple graph with no isolated vertices (but possibly disconnected), and let A be its adjacency matrix. Let $\mathrm{spec}(A)$ denote the set of distinct eigenvalues of A. For an eigenvalue μ of A, let U_μ be its eigenspace, let n_μ be its multiplicity, and let $u_{\mu,1}, \ldots, u_{\mu,n_\mu}$ be the orthonormal basis of U_μ. From the spectral decomposition, adjacency matrix A can be represented as

$$A = \sum_{\mu \in \mathrm{spec}(A)} \sum_{k=1}^{n_\mu} \mu \, u_{\mu,k} u_{\mu,k}^T,$$

while it holds that

$$I = \sum_{\mu \in \mathrm{spec}(A)} \sum_{k=1}^{n_\mu} u_{\mu,k} u_{\mu,k}^T, \tag{2.32}$$

due to orthonormality of chosen eigenvectors.

Note that for an eigenvector u of A with an eigenvalue μ and any family $\beta_1, \beta_2, \ldots, \beta_\tau$ of real numbers, we have

$$\prod_{t=1}^{\tau} (A - \beta_t I) u = \prod_{t=1}^{\tau} (\mu - \beta_t) u. \tag{2.33}$$

In addition, the rank of the matrix $\prod_{t=1}^{\tau} (A - \beta_t I)$ coincides with the sum of the dimensions of the eigenspaces U_μ, summed over all eigenvalues μ not in $\{\beta_1, \ldots, \beta_\tau\}$:

$$\mathrm{rk}\left(\prod_{t=1}^{\tau} (A - \beta_t I) \right) = \sum_{\mu \in \mathrm{spec}(A) \backslash \{\beta_1, \ldots, \beta_\tau\}} \dim U_\mu.$$

From (2.32) we now have

$$\prod_{t=1}^{\tau} (A - \beta_t I) = \prod_{t=1}^{\tau} (A - \beta_t I) I$$

$$= \prod_{t=1}^{\tau} (A - \beta_t I) \sum_{\mu \in \mathrm{spec}(A)} \sum_{k=1}^{n_\mu} u_{\mu,k} u_{\mu,k}^T$$

$$= \sum_{\mu \in \mathrm{spec}(A)} \sum_{k=1}^{n_\mu} \prod_{t=1}^{\tau} (A - \beta_t I) u_{\mu,k} u_{\mu,k}^T$$

$$= \sum_{\mu \in \mathrm{spec}(A)} \sum_{k=1}^{n_\mu} \prod_{t=1}^{\tau} (\mu - \beta_t) u_{\mu,k} u_{\mu,k}^T \qquad \text{by (2.33)}$$

$$= \sum_{\mu \in \mathrm{spec}(A) \setminus \{\beta_1, \dots, \beta_\tau\}} \sum_{k=1}^{n_\mu} \prod_{t=1}^{\tau} (\mu - \beta_t) u_{\mu,k} u_{\mu,k}^T. \qquad (2.34)$$

If we now specialize to the case

$$\mathrm{spec}(A) = \{\mu_1 > \mu_2 > \cdots > \mu_p\}, \quad \mu = \mu_1,$$
$$\{\beta_1, \dots, \beta_\tau\} = \{\mu_2, \dots, \mu_p\},$$

then all summands in (2.34) vanish, except the one for μ_1, so that

$$\prod_{i=2}^{p} (A - \mu_i I) = \sum_{k=1}^{n_{\mu_1}} \prod_{i=2}^{p} (\mu_1 - \mu_i) u_{\mu_1,k} u_{\mu_1,k}^T, \qquad (2.35)$$

and the rank of $\prod_{i=2}^{p}(A - \mu_i I)$ is equal to the multiplicity of the spectral radius μ_1.

The proof of Hoffman's characterization is now evident: if (2.31) holds, then by recalling that $J = jj^T$, we have

$$\prod_{i=2}^{p} (A - \mu_i I) = \prod_{i=2}^{p} (\mu_1 - \mu_i) \frac{j}{\sqrt{n}} \frac{j^T}{\sqrt{n}},$$

showing that the spectral radius is a simple eigenvalue, whose eigenspace is generated by $j/\sqrt{n}$, i.e., that G is a connected, regular graph. The other direction follows directly from (2.35).

Next, a harmonic graph $G = (V, E)$ is characterized by the fact that Aj is the eigenvector corresponding to the spectral radius, so that from (2.35) we have

$$\prod_{i=2}^{p} (A - \mu_i I) = \prod_{i=2}^{p} (\mu_1 - \mu_i) \frac{(Aj)(Aj)^T}{||Aj||^2}.$$

Now, from

$$||Aj||^2 = (Aj)^T (Aj) = j^T A^T A j = j^T A^2 j = \mu_1 j^T A j = 2 \mu_1 |E|,$$

we obtain

Theorem 2.15 ([52]). *If $G = (V, E)$ is a graph without isolated vertices, then G is a connected, harmonic graph if and only if*

$$2\mu_1 |E| \prod_{i=2}^{p}(A - \mu_i I) = \prod_{i=2}^{p}(\mu_1 - \mu_i)AJA.$$

Finally, a μ_1^2-semiharmonic graph $G = (V, E)$ is characterized by the fact that $A^2 j + \mu_1 A j$ is the eigenvector corresponding to the spectral radius, so that from (2.35) we have

$$\prod_{i=2}^{p}(A - \mu_i I) = \prod_{i=2}^{p}(\mu_1 - \mu_i)\frac{(A^2 j + \mu_1 A j)(A^2 j + \mu_1 A j)^T}{||A^2 j + \mu_1 A j||^2}.$$

Now

$$(A^2 j + \mu_1 A j)(A^2 j + \mu_1 A j)^T = A(AJA + \mu_1 AJ + \mu_1 JA + \mu_1^2 J)A,$$

while

$$
\begin{aligned}
||A^2 j + \mu_1 A j||^2 &= (A^2 j + \mu_1 A j)^T (A^2 j + \mu_1 A j) \\
&= j^T A^4 j + 2\mu_1 j^T A^3 j + \mu_1^2 j^T A^2 j \\
&= \mu_1^2 j^T A^2 j + 2\mu_1^3 j^T A j + \mu_1^2 j^T A^2 j \qquad \text{(by } A^3 j = \mu_1^2 A j) \\
&= 2\mu_1^2 (j^T A^2 j + \mu_1 j^T A j).
\end{aligned}
$$

The expression $j^T A^2 j$ represents the number of all walks of length 2 in G, which, if the walks are counted by their middle vertex, is equal to $M_1 = \sum_{u \in V} \deg_u^2$. Hence,

Theorem 2.16 ([52]). *If $G = (V, E)$ is a graph without isolated vertices, then G is a connected, μ_1^2-semiharmonic graph if and only if*

$$2\mu_1^2(M_1 + 2\mu_1 |E|) \prod_{i=2}^{p}(A - \mu_i I) = \prod_{i=2}^{p}(\mu_1 - \mu_i)A(AJA + \mu_1 AJ + \mu_1 JA + \mu_1^2 J)A.$$

CHAPTER 3

Spectral Radius of Particular Types of Graphs

We continue our journey by showcasing results on the spectral radius of graphs belonging to particular graph classes. Section 3.1 deals with the question of how close to the maximum vertex degree can the spectral radius be if a graph is not regular. Section 3.2 then shows how to construct the graph with the maximum spectral radius given the sequence of its vertex degrees. Section 3.3 contains recent results on the graphs with a few edges: trees and planar graphs. Finally, Section 3.4 gives results on extremal spectral radii of complete multipartite graphs, together with an interesting conjecture of Delorme on the concavity of spectral radius over a simplex of graph parameters.

3.1 NONREGULAR GRAPHS

We know from (1.5) and Section 2.5 that if G is a regular graph, then $\lambda_1 = \Delta$, while if G is not a regular graph, then $\lambda_1 < \Delta$. The natural question arises: if G is not regular, how close can its spectral radius λ_1 be to Δ? This question, first posed by Stevanović [141], started a whole little industry of papers on the estimates of $\Delta - \lambda_1$ in nonregular graphs.

Let $G = (V, E)$ be a nonregular graph and let λ_1 and x be its spectral radius and the principal eigenvector. From the Rayleigh quotient we have

$$\lambda_1 = \frac{x^T A x}{x^T x} = 2 \sum_{uv \in E} x_u x_v,$$

so that

$$
\begin{aligned}
\Delta - \lambda_1 &= \sum_{u \in V} \Delta x_u^2 - 2 \sum_{uv \in E} x_u x_v \\
&= \sum_{u \in V} (\Delta - \deg_u) x_u^2 + \sum_{u \in V} \deg_u x_u^2 - 2 \sum_{uv \in E} x_u x_v \\
&= \sum_{u \in V} (\Delta - \deg_u) x_u^2 + \sum_{uv \in E} \left(x_u^2 + x_v^2 - 2 x_u x_v \right) \\
&= \sum_{u \in V} (\Delta - \deg_u) x_u^2 + \sum_{uv \in E} (x_u - x_v)^2.
\end{aligned}
$$

Spectral Radius of Graphs. http://dx.doi.org/10.1016/B978-0-12-802068-5.00003-8

Let a be the vertex having the maximum principal eigenvector component x_{max}, and let b be the vertex having the minimum principal eigenvector component x_{min}. Further, let $P: a = w_0, w_1, \ldots, w_p = b$ be the shortest path between a and b in G. As P is a subgraph of G, we get from the Cauchy-Schwarz inequality that

$$\sum_{uv \in E} (x_u - x_v)^2 \geq \sum_{i=0}^{p-1} (x_{w_i} - x_{w_{i+1}})^2 \geq \frac{1}{p} \left(\sum_{i=0}^{k-1} (x_{w_i} - x_{w_{i+1}}) \right)^2$$
$$= \frac{1}{p} (x_{max} - x_{min})^2.$$

If we also observe that $p \leq D$, where D is the diameter of G, then we have

$$\Delta - \lambda_1 \geq \sum_{u \in V} (\Delta - \deg_u) x_u^2 + \frac{1}{D} (x_{max} - x_{min})^2. \tag{3.1}$$

Different authors have provided different endings to the above argument.

Theorem 3.1 ([141]). $\Delta - \lambda_1 > \dfrac{1}{2n(n\Delta - 1)\Delta^2}.$

Proof. Stevanović has avoided estimating x_{max} and x_{min} by dividing the proof in two cases. Let $c > 0$ be a positive number. Since $\sum_{u \in V} x_u^2 = 1$, we necessarily have

$$x_{max} \geq \frac{1}{\sqrt{n}} \geq x_{min}.$$

Case I: If $x_{max} \geq \frac{1}{\sqrt{n}} + c$ or $\frac{1}{\sqrt{n}} - c \geq x_{min}$, then $(x_{max} - x_{min})^2 \geq c^2$.

Case II: If $\frac{1}{\sqrt{n}} + c > x_{max} > x_{min} > \frac{1}{\sqrt{n}} - c$, then $x_u \in \left(\frac{1}{\sqrt{n}} - c, \frac{1}{\sqrt{n}} + c \right)$ for each $u \in V$. If we now choose $s \in V$ with $\deg_s \leq \Delta - 1$, then from the eigenvalue equation we get

$$\lambda_1 \left(\frac{1}{\sqrt{n}} - c \right) < \lambda_1 x_s = \sum_{t \in N_s} x_t < (\Delta - 1) \left(\frac{1}{\sqrt{n}} + c \right),$$

which yields

$$\lambda_1 < (\Delta - 1) \frac{1 + c\sqrt{n}}{1 - c\sqrt{n}}.$$

It now suffices to put $c = \frac{1}{2\Delta\sqrt{n}}$ to finish the proof. The first summand on the right-hand side of (3.1) is nonnegative, as $\Delta \geq \deg_u$ for each $u \in V$. Using also that $D \leq |E| \leq \frac{n\Delta-1}{2}$ (as G is not regular), we get in Case I that $\lambda_1 < \Delta - \frac{1}{2n(n\Delta-1)\Delta^2}$, while in Case II we get that $\lambda_1 < \Delta - \frac{1}{2\Delta-1} < \Delta - \frac{1}{2n(n\Delta-1)\Delta^2}$. $\qquad\square$

Theorem 3.2 ([168]). $\Delta - \lambda_1 > \dfrac{(\sqrt{\Delta} - \sqrt{\delta})^2}{nD\Delta}$.

Proof. Zhang relied on the Ostrowski inequality (2.15):

$$\frac{x_{\max}}{x_{\min}} \geq \sqrt{\frac{\Delta}{\delta}},$$

where δ is the minimum vertex degree, and $x_{\max} > \frac{1}{\sqrt{n}}$ (as the graph is not regular), to deduce that

$$x_{\max} - x_{\min} \geq \frac{\sqrt{\Delta} - \sqrt{\delta}}{\sqrt{\Delta}} x_{\max} > \frac{\sqrt{\Delta} - \sqrt{\delta}}{\sqrt{n\Delta}}.$$

Putting this back into (3.1) yields

$$\Delta - \lambda_1 > \frac{(\sqrt{\Delta} - \sqrt{\delta})^2}{nD\Delta}.$$

$\qquad\square$

Theorem 3.3 ([96]). $\Delta - \lambda_1 \geq \dfrac{\Delta + 1}{n(3n + 2\Delta - 4)}$.

Proof. Liu et al. have observed that

$$\sum_{u \in V}(\Delta - \deg_u)x_u^2 \geq x_{\min}^2,$$

as at least one vertex has degree strictly smaller than Δ. Hence, from (3.1) one obtains

$$\Delta - \lambda_1 \geq x_{\min}^2 + \frac{1}{D}(x_{\max} - x_{\min})^2 \geq \frac{x_{\max}^2}{D+1}.$$

The second inequality follows by considering $x_{\min}^2 + \frac{1}{D}(x_{\max} - x_{\min})^2$ as a quadratic function in $x_{\min}$, whose minimum value is obtained for

$x_{\min} = \frac{x_{\max}}{D}$, and a simple inequality $\frac{1}{D^2} + \frac{(D-1)^2}{D^3} > \frac{1}{D+1}$. To finish the proof, Liu et al. relied on $x_{\max} > \frac{1}{\sqrt{n}}$ and the following bound on the diameter of a nonregular graph with the maximum spectral radius.

Lemma 3.1. *Let G be a connected, nonregular graph with n vertices and the maximum vertex degree Δ. If G has the maximum spectral radius among all connected graphs with n vertices and the maximum vertex degree Δ, then for its diameter D holds*

$$D \le \frac{3n + \Delta - 5}{\Delta + 1}$$

with equality if and only if $\Delta = 2$ and G is a path.   □

Later, Liu et al. [91, 95] slightly improved the above bound on the diameter to

$$D \le \frac{3n + \Delta - 8}{\Delta + 1}.$$

They have also shown that in the graph with the maximum spectral radius among all connected graphs with n vertices and the maximum vertex degree Δ, vertices of degree less than Δ necessarily induce a complete subgraph.

Theorem 3.4 ([35]). $\Delta - \lambda_1 > \dfrac{n\Delta - 2m}{n((n\Delta - 2m)D + 1)}.$

Proof. Cioabă et al. [35] noticed that the first summand on the right-hand side of (3.1) may be bounded as

$$\sum_{u \in V} (\Delta - \deg_u) x_u^2 \ge \sum_{u \in V} (\Delta - \deg_u) x_{\min}^2 = (n\Delta - 2m) x_{\min}^2,$$

so that (3.1) becomes

$$\Delta - \lambda_1 \ge (n\Delta - 2m) x_{\min}^2 + \frac{1}{D}(x_{\max} - x_{\min})^2.$$

If we now consider $(n\Delta - 2m)x_{\min}^2 + \frac{1}{D}(x_{\max} - x_{\min})^2$ as a quadratic function in $x_{\min}$, it reaches its minimum value when $x_{\min} = \frac{x_{\max}}{(n\Delta - 2m)D + 1}$. Together with $x_{\max}^2 \ge \frac{1}{n}$, we get

$$\Delta - \lambda_1 > \frac{n\Delta - 2m}{n((n\Delta - 2m)D + 1)}.$$

□

Shi [134] has obtained further improvements by applying the Cauchy-Schwarz inequality to arbitrary terminal segments of the shortest path $P: a = w_0, \ldots, w_p = b$ from the vertex a with the maximum principal eigenvector component x_{max} to the vertex b with the minimum principal eigenvector component x_{min}:

$$\sum_{i=j}^{p-1}(x_{w_i} - x_{w_{i+1}})^2 \geq \frac{1}{p-j}\left(\sum_{i=j}^{p-1}(x_{w_i} - x_{w_{i+1}})\right)^2 = \frac{1}{p-j}(x_{w_j} - x_{min})^2,$$

and then dividing the proof in several cases, depending on whether x_{min}^2, $\sum_{v \in N_b} x_v^2$, or $x_{w_j}^2$, $j = 0, \ldots, p - 1$, is bounded from below by suitable expressions. As a result, Shi has obtained

Theorem 3.5 ([134]). *If G is a connected, nonregular graph with n vertices, the diameter D, the maximum vertex degree Δ, and the minimum vertex degree δ, then*

$$\Delta - \lambda_1 > \left[(n - \delta)D + \frac{1}{\Delta - \frac{2m}{n}} - \binom{D}{2}\right]^{-1}.$$

A different approach was taken by Cioabă [31] in order to improve the bound $\Delta - \lambda_1 > \frac{1}{n(D+1)}$, that is implicitly contained in the proof of Theorem 3.3.

Theorem 3.6 ([31]). $\Delta - \lambda > \frac{1}{nD}$.

Proof. Let a be the vertex with the principal eigenvector component x_{max}. If $\deg_a \leq \Delta$, then

$$\lambda_1 x_{max} = \sum_{v \in N_a} x_v \leq (\Delta - 1)x_{max},$$

which implies $\Delta - \lambda_1 \geq 1 > \frac{1}{nD}$.

Assume, therefore, that $\deg_a = \Delta$. The proof is split in two cases depending on the number of vertices in G whose degree is less than Δ.

Case I: *G contains at least two vertices whose degree is less than Δ.*

Let u and v be two vertices with $\deg_u, \deg_v < \Delta$. Let $P_u : u = i_0, i_1, \ldots, i_r = a$ be a shortest path from u to a, and let $P_v : v = j_0, j_1, \ldots, j_s = a$ be a shortest path from v to a. Let t_u, $0 \le t \le r$, be the smallest index such that i_{t_u} belongs to P_v as well, and let t_v be such that $i_{t_u} = j_{t_v}$.

If $t_u = 0$, then $r < s \le D$, so that $r \le D - 1$ and similarly to (3.1) we have

$$\Delta - \lambda_1 = \sum_{k \in V} (\Delta - \deg_k) x_k^2 + \sum_{kl \in E} (x_k - x_l)^2$$

$$\ge x_u^2 + \sum_{q=0}^{r-1} (x_{i_{q+1}} - x_{i_q})^2$$

$$\ge \frac{1}{r+1} \left(x_u + \sum_{q=0}^{r-1} (x_{i_{q+1}} - x_{i_q}) \right)^2$$

$$= \frac{1}{r+1} x_{i_r}^2 = \frac{1}{r+1} x_{\max}^2 > \frac{1}{nD}.$$

If $t_u \ge 1$, then we may assume without loss of generality that $t_u \ge t_v$. From the Cauchy-Schwarz inequality follows that

$$\Delta - \lambda_1 = \sum_{k \in V} (\Delta - \deg_k) x_k^2 + \sum_{kl \in E} (x_k - x_l)^2$$

$$\ge x_u^2 + x_v^2 + \sum_{q=0}^{r-1} (x_{i_{q+1}} - x_{i_q})^2 + \sum_{q'=0}^{t_v-1} (x_{j_{q'+1}} - x_{j_{q'}})^2$$

$$= \left(x_u^2 + \sum_{q=0}^{t_u-1} (x_{i_{q+1}} - x_{i_q})^2 \right) + \left(x_v^2 + \sum_{q'=0}^{t_v-1} (x_{j_{q'+1}} - x_{j_{q'}})^2 \right)$$

$$+ \sum_{q=t_u}^{r-1} (x_{i_{q+1}} - x_{i_q})^2$$

$$\ge \frac{x_{i_{t_u}}^2}{t_u + 1} + \frac{x_{j_{t_v}}^2}{t_v + 1} + \frac{(x_{i_{t_u}} - x_{\max})^2}{r - t_u}$$

$$\geq \frac{2x_{i_{t_u}}^2}{t_u + 1} + \frac{(x_{\max} - x_{i_{t_u}})^2}{r - t_u}.$$

The last expression is a quadratic function in $x_{i_{t_u}}$, which attains its minimum when $x_{i_{t_u}} = \frac{t_u+1}{2r-t_u+1}x_{\max}$. This implies that

$$\Delta - \lambda_1 > \frac{2x_{\max}^2}{2r - t_u + 1} \geq \frac{x_{\max}^2}{r} \geq \frac{1}{nD}.$$

Case II: *G contains exactly one vertex whose degree is less than Δ.*

Let b be the vertex with the principal eigenvector component $x_{\min}$. Then $\deg_b < \Delta$ follows from

$$\Delta x_{\min} > \lambda_1 x_{\min} = \sum_{j \in N_b} x_j \geq \deg_b x_{\min}.$$

Let $\gamma = \frac{x_{\max}}{x_{\min}}$. If $\gamma \leq D$, then by summing the eigenvalue equation $\lambda_1 x_i = \sum_{j \in N_i} x_j$ for all $i \in V$, we get

$$\Delta - \lambda_1 = \frac{(\Delta - \deg_b)x_{\min}}{\sum_{i \in V} x_i} > \frac{x_{\min}}{nx_{\max}} = \frac{1}{n\gamma} \geq \frac{1}{nD}.$$

We assume, therefore, that $\gamma > D$. We may also assume that $d(a, b) = D$, as otherwise we can easily finish the proof by applying an argument similar to the one of the previous case.

We claim that there exists $j \in N_a$ such that $x_j < \frac{1}{\sqrt{n}}$. Otherwise, let $j \in N_a$ be such that $d(j, b) = D - 1$. Then applying the argument from the previous case gives

$$\Delta - \lambda_1 > \frac{x_j^2}{D} > \frac{1}{nD}.$$

From $x_j < \frac{1}{\sqrt{n}}$ and $j \in N_a$ we have

$$\lambda_1 x_{\max} = \sum_{i \in N_a} x_i < (\Delta - 1)x_{\max} + \frac{1}{\sqrt{n}},$$

which implies

$$\Delta - \lambda_1 > 1 - \frac{1}{x_{\max}\sqrt{n}}.$$

If the right-hand side is at least $\frac{1}{nD}$, then we are done. Otherwise, $1 - \frac{1}{x_{\max}\sqrt{n}} < \frac{1}{nD}$ implies

$$x_{\max} < \frac{D\sqrt{n}}{nD - 1}. \tag{3.2}$$

Now from

$$1 = \sum_{i \in V} x_i^2 \leq (n-1)x_{\max}^2 + x_{\min}^2,$$

we get

$$x_{\min}^2 \geq 1 - \frac{(n-1)nD^2}{(nD-1)^2} = \frac{(nD-1)^2 - (n-1)nD^2}{(nD-1)^2} = \frac{(D^2 - 2D)n + 1}{(nD-1)^2}. \tag{3.3}$$

Assume now that $D \geq 3$. From (3.2) and (3.3) we get that

$$\gamma^2 = \frac{x_{\max}^2}{x_{\min}^2} < \frac{D^2 n}{(D^2 - 2D)n + 1} < D^2,$$

which is in contradiction with the earlier assumption $\gamma > D$. Therefore, we have that $D = 2$ must hold.

Looking at the square of the adjacency matrix of G we get

$$\lambda_1 x_{\max} \leq (\Delta^2 - 1)x_{\max} + x_{\min}, \tag{3.4}$$

which implies

$$\lambda_1^2 \leq \Delta^2 - 1 + \frac{1}{\gamma} < \Delta^2 - \frac{1}{2}, \tag{3.5}$$

since $\gamma > D = 2$. Note that the inequality (3.4) holds as there exists at least one walk of length two between a and b. From (3.5) we further have

$$\lambda_1 < \sqrt{\Delta^2 - \frac{1}{2}} < \Delta - \frac{1}{4\Delta}.$$

If $n \geq 2\Delta$, then we are done, as then $4\Delta \leq nD$.

Assume, therefore, that $n < 2\Delta$. In that case the vertex a has at least two neighbors that are also adjacent to the vertex b. We deduce that

$$\lambda_1^2 x_{\max} \leq (\Delta^2 - 2)x_{\max} + 2x_{\min}$$

which implies that

$$\lambda_1^2 \leq \Delta^2 - 2 + \frac{2}{\gamma} \geq \Delta^2 - 1.$$

Thus,

$$\lambda_1 \leq \sqrt{\Delta^2 - 1} < \Delta - \frac{1}{2\Delta} < \Delta - \frac{1}{nD},$$

which completes the proof.   □

Let $\lambda_1(n, \Delta)$ be the maximum spectral radius of nonregular graphs with n vertices and the maximum vertex degree Δ. Another interesting question is on the order of $\Delta - \lambda_1(n, \Delta)$ when n is let to tend to infinity. Cioabă et al. [35] constructed an infinite family of nonregular graphs with the maximum vertex degree Δ such that $\Delta - \lambda \leq \frac{4\pi^2}{nD}$, while Liu et al. [96] constructed another family such that $\Delta - \lambda_1 \leq \frac{3\pi^2}{nD}$. In both cases, the diameter D is $O(\frac{n}{\Delta})$. Liu et al. went even further to propose the following

Conjecture 3.1 ([96]). *For each fixed Δ*

$$\lim_{n \to \infty} \frac{n^2}{\Delta - 1}(\Delta - \lambda_1(n, \Delta)) = \pi^2.$$

Immediate corollary of Theorem 3.6 is the following result, which was independently proved by Nikiforov [112].

Theorem 3.7 ([31, 112]). *If H is a proper subgraph of a Δ-regular graph G with diameter D, then*

$$\Delta - \lambda_1(H) > \frac{1}{nD}.$$

Combining Nikiforov's approach from [112] with his approach for the proof of Theorem 3.5, Shi obtained the following

Theorem 3.8 ([134]). *If H is a proper subgraph of a Δ-regular graph G with diameter D, then*

$$\Delta - \lambda_1(H) > \begin{cases} \left[(n-\Delta)D + \Delta - 1 - \binom{D-1}{2}\right]^{-1}, & \text{for } \Delta \leq \frac{n}{2} + 2, \\ \frac{2}{3(n-2)}, & \text{for } \Delta > \frac{n}{2} + 2. \end{cases}$$

Another part of (1.5) is the inequality $\lambda_1 \geq \frac{2m}{n}$, with equality if and only if the connected graph is regular. Nikiforov [111] has studied the difference $\lambda_1 - \frac{2m}{n}$ for nonregular graphs. Call a graph *subregular* if $\Delta - \delta = 1$ and all vertices, except one, has the same degree.

Theorem 3.9 ([111]). *If G is not regular and not subregular, then*

$$\lambda_1 - \frac{2m}{n} > \frac{1}{2m + 2n}.$$

If G is subregular, then

$$\lambda_1 - \frac{2m}{n} > \frac{1}{n\Delta + 2n}. \tag{3.6}$$

Inequality (3.6) was proved earlier for nonregular graphs by Cioabă and Gregory [34]. Nikiforov's proof in [111] is based on the Hofmeister's inequality [81]

$$\lambda_1^2 \geq \frac{\sum_{u \in V} \deg_u^2}{n},$$

case-by-case analysis, and lots of arithmetic, so we will skip it here.

3.2 GRAPHS WITH A GIVEN DEGREE SEQUENCE

Motivated by the still partly open problem of characterizing the graph with the maximum spectral radius among connected graphs with given numbers of vertices and edges (see the last part of Section 2.4), Biyikoğlu and Leydold [17] have considered the maximum spectral radius of connected graphs with a given degree sequence π.

Following [17], let us introduce an ordering of the vertices of a graph G by means of the breadth-first search: select an arbitrary vertex of G and denote it by v_0; then select all neighbors of v_0, sorted by decreasing degrees, and denote them by $v_1, \ldots, v_{\deg_{v_0}}$; then select all neighbors of v_1 that do not already appear in the list, and continue iteratively with $v_2, v_3, \ldots$ until all vertices of G are processed. In this way, we build layers where each vertex v in the layer i has distance i to the root v_0, and it is adjacent to some of the vertices in layer $i - 1$. We call the parent of v the vertex w with the smallest index among all vertices adjacent to v in layer $i - 1$, and also say that v is a child of w. Vertices of G, denoted in this way, can be drawn along a spiral, hence the name spiral-like ordering [124] or spiral-like disposition [10].

Definition 3.1 ([17]). Let $G = (V, E)$ be a connected graph with root v_0. Then a well-ordering $\prec$ of the vertices is called the *breadth-first search ordering with decreasing degrees* (BFD-ordering for short) if the following holds for all vertices $v, w \in V$:

(B1)   if $w_1 \prec w_2$ then $z_1 \prec z_2$ for all children z_1 of w_1 and z_2 of w_2;
(B2)   if $v \prec u$ then $\deg_v \geq \deg_u$.

A connected graph that has a BFD-ordering of its vertices will be called a BFD-graph. Each connected graph has an ordering that satisfies (B1), as given by the spiral-like ordering starting with any of its vertices. However, not all graphs have an ordering that also satisfies (B2).

For a given degree sequence $\pi : d_0 \geq d_1 \geq \cdots \geq d_{n-1}$ of positive integers, let

$$C_\pi = \{G : G \text{ is a connected graph with degree sequence } \pi\}.$$

The following two lemmas will be useful in later proofs.

Lemma 3.2 ([17]). *Let x be the principal eigenvector of a graph G in C_π. If G contains two vertices such that $\deg_u > \deg_v$ and $x_u < x_v$, then G cannot have the maximum spectral radius in C_π.*

Proof. Let $\deg_u - \deg_v = c$ and assume $x_u < x_v$. Then there are at least c neighbors $w_1, \ldots, w_c$ of u that are not adjacent to v. If we create a new graph G' by replacing the edges $w_1 u, \ldots, w_c u$ by the edges $w_1 v, \ldots, w_c v$, then G' has the same degree sequence π. The neighbors $w_1, \ldots, w_c$ can be chosen such that G' remains connected, as either u and v have a common neighbor, or u and v are adjacent, or we can select any among the neighbors of u. By Lemma 1.2 we then have $\lambda_1(G') > \lambda_1(G)$.   □

Lemma 3.3 ([17]). *Let x be the principal eigenvector of a graph G in C_π. Let $vu \in E(G)$ and $vt \notin E(G)$ with $x_u < x_t \leq x_v$. If $x_v \geq x_w$ for all neighbors w of t, then G cannot have the maximum spectral radius in C_π.*

Proof. Assume that such vertices exist. Construct a new graph $G' = (V, E')$ with the same degree sequence π by replacing the edge vu by vt and the edge tw by uw for some neighbor w of t. Then $\lambda_1(G') > \lambda_1(G)$ by Lemma 1.3.

It remains to show that we can choose the neighbor w such that G' is connected. First, notice that there must be a neighbor p of t that is not adjacent to u, since otherwise

$$\sum_{w \in N_t} x_w \leq \sum_{w \in N_u} x_w,$$

and then by the eigenvalue equation

$$\lambda_1(G)x_t \leq \lambda_1(G)x_u,$$

contradictory to the assumption $x_t > x_u$.

Furthermore, t must have at least two neighbors, as otherwise for the unique neighbor w of t we would have

$$x_w = \lambda_1 x_t > \lambda_1 x_u \geq x_v,$$

contradictory to the assumption $x_v \geq x_w$.

Since G is connected, there exists a path $P_{vt} \colon v = z_0, \ldots, z_{k-1}, z_k = t$ from v to t. Then there are four cases:

1) If $vu \notin P_{vt}$ and $uz_{k-1} \notin E(G)$, then we set $w = z_{k-1}$.
2) If $vu \notin P_{vt}$ and $uz_{k-1} \in E(G)$, then we set w to one of the neighbors of t that are not adjacent to u.
3) If $vu \in P_{vt}$ and all neighbors of t, other than z_{k-1}, are adjacent to u, then z_{k-1} cannot be adjacent to u we set $w = z_{k-1}$.
4) If $vu \in P_{vt}$ and there exists a neighbor p of t, $p \neq z_{k-1}$, with $up \notin E(G)$, then we set $w = p$.

In either case, G' is connected, so that $G' \in \mathcal{C}_\pi$.   □

Theorem 3.10 ([17]). *Let G have the maximum spectral radius in $\mathcal{C}_\pi$. Then there exists a BFD-ordering of its vertices that is consistent with its principal eigenvector x in a sense that $x_u > x_v$ implies $u \prec v$, hence $\deg_u \geq \deg_v$.*

Proof. Let x be the principal eigenvector of G. Create an ordering $\prec$ of the vertices of G by breadth-first search as follows: choose the vertex with the maximum principal eigenvector component $x_{\max}$ as v_0; append all neighbors $v_1, \ldots, v_{\deg_{v_0}}$ of v_0; these neighbors are ordered such that $u \prec v$ whenever $\deg_u > \deg_v$, or $\deg_u = \deg_v$ and $x_u > x_v$; then continue iteratively with the vertices $v_1, v_2, \ldots$ until all vertices of G are processed. Notice that the property (B1) holds for this ordering.

We first show that $u \prec v$ implies $x_u \geq x_v$ for all $u, v \in V(G)$. Suppose that there exist two vertices v_i and v_j with $v_i \prec v_j$, but $x_{v_i} < x_{v_j}$. Notice that v_i cannot be the root v_0. Let w_i and w_j be the parents of v_i and v_j, respectively. By construction there are two cases: (i) $w_i = w_j$, or (ii) $w_i \prec w_j$. For case (i) we have $\deg_{v_i} \geq \deg_{v_j}$ by construction and $\deg_{v_i} \leq \deg_{v_j}$ by Lemma 3.2, and thus $\deg_{v_i} = \deg_{v_j}$. However, then we have $v_i \succ v_j$ by the definition of the ordering, since $x_{v_j} > x_{v_i}$, a contradiction.

For case (ii) assume that v_j is maximal, i.e., for any other vertex u with this property we have $x_u \leq x_{v_j}$. Let $v_i(\prec v_j)$ be the first vertex, according to the ordering $\prec$, with $x_{v_i} < x_{v_j}$. Hence $x_u \geq x_{v_j}$ for each $u \prec v_i$ and we have $x_{w_i} \geq x_{v_j} > x_{v_i}$. Note that v_j cannot be adjacent either to w_i or to v_0, as we would then have case (i). Thus $x_{w_i} \geq x_{u_j}$ for all neighbors u_j of v_j, as otherwise v_j was not maximal. Hence G cannot have the maximum spectral radius by Lemma 3.3, a contradiction.

At last, we have to show that the property (B2) holds. This, however, follows immediately from Lemma 3.2.   □

It is possible that there exist two BFD-graphs for a given degree sequence π. However, when restricted to degree sequences of trees, exactly one BFD-graph exists.

Theorem 3.11 ([17]). *A tree T with degree sequence π has the maximum spectral radius in C_π if and only if T is a BFD-graph. T is then, up to isomorphism, determined uniquely by π.*

Proof. The necessary condition is an immediate corollary of Theorem 3.10. In order to show that two BFD-trees T and T' in C_π are isomorphic, we use a function ϕ that maps the vertex v_i in the i-th position in the BFD-ordering of T to the vertex w_i in the ith position in the BFD-ordering of T'. By properties (B1) and (B2), ϕ is an isomorphism, as v_i and w_i have the same degree and the images of neighbors of v_i in the next layer in T are exactly the neighbors of w_i in the next layer in T'. The latter can be seen by considering all vertices of T in the reverse BFD-ordering.   □

Belardo et al. [10] extended the previous theorem to describe a unicyclic graph with the degree sequence and the maximum spectral radius given.

Theorem 3.12 ([10]). *Let π be a degree sequence of a unicyclic graph, and let U have the maximum spectral radius in C_π. Let v_1, v_2, and v_3 be the vertices of U having the largest degrees. Then v_1, v_2, and v_3 form a triangle, while the remaining vertices of U appear in spiral-like ordering, starting from the vertices adjacent to v_1.*

Belardo et al. [10] have further conjectured that for a given k and for any degree sequence $\pi : d_0 \geq \cdots \geq d_{n-1}$ with $\sum_{i=0}^{n-1} d_i = 2(n+k)$, the graph G with the maximum spectral radius in C_π may be constructed by the following procedure. Assume that $v_0, v_1, \ldots, v_{n-1}$ are the vertices of G, ordered by their degrees. Also, H_j is an intermediate graph such that $G = H_{n-1}$ at the end of the algorithm.

1) Set $j = 0$ and $H_0 = nK_1$.
2) Increase j by one, and add edges to H_j by joining v_j to v_i for i, $i = 0, \ldots, j - 1$, whenever the degree of v_i in H_{j-1} is less than d_i, provided also that the graph obtained so far is not $(k + 1)$-cyclic; otherwise, if all vertices $v_0, v_1, \ldots, v_{j-1}$ have appropriate degrees in H_{j-1}, then prior to adding new edges, delete the last edge added in the $(j - 1)$-st step, and proceed with the jth step, or go to 3) if H_j is $(k + 1)$-cyclic. Otherwise, repeat (2).
3) If $j = n - 1$, then stop and output $G = H_{n-1}$. Increase j by one, and join v_j only to the vertex v_i of the maximum degree in H_{j-1}, whose degree in H_{j-1} is less than d_i. Repeat (3).

It is also possible to compare maximum spectral radii in classes corresponding to distinct degree sequences under suitable conditions. For two distinct degree sequences $\pi : d_0 \geq \cdots \geq d_{n-1}$ and $\pi' : d_0' \geq \cdots \geq d_{n-1}'$ we say that π is majorized by π' and write $\pi \lhd \pi'$ if $\sum_{i=0}^{j} d_i \leq \sum_{i=0}^{j} d_i'$ for all $j = 0, \ldots, n - 1$.

Theorem 3.13 ([17]). *Let π and π' be two distinct degree sequences of trees with $\pi \lhd \pi'$. If T and T' are the trees with the maximum spectral radius in C_π and $C_{\pi'}$, respectively, then $\lambda_1(T) < \lambda_1(T')$.*

Proof. Let $\pi : d_0 \geq d_1 \geq \ldots d_{n-1}$ and $\pi' : d_0' \geq d_1' \geq \cdots \geq d_{n-1}'$. If T has the maximum spectral radius in C_π, then by Theorem 3.11, T has a BFD-ordering that is consistent with its principal eigenvector x, i.e., $x_u > x_v$ implies $u \prec v$.

Assume firstly that π and π' differ only in two positions k and l with $d'_k = d_k + 1$ and $d'_l = d_l - 1$ (hence $k < l$ and $d_k \geq d_l > 1$). Let v_k and v_l be the corresponding vertices in T. Without loss of generality, we assume that $x_{v_k} \geq x_{v_l}$. Since T is a tree and $\deg_{v_l} \geq 2$, there exists a neighbor w of v_l in the next layer that is not adjacent to v_k. Thus we can rotate the edge $v_l w$ to $v_k w$ and obtain a new tree T' with degree sequence π' and $\lambda_1(T') > \lambda_1(T)$ by Lemma 1.1.

For two tree degree sequences $\pi \lhd \pi'$ we can find a sequence of tree degree sequences $\pi = \pi_0 \lhd \pi_1 \lhd \cdots \lhd \pi_k = \pi'$, where π_{i-1} and π_i, $i = 1, \ldots, k$, differ in two positions only, as described by the following procedure. For π_{i-1} let j be the first position in which π_{i-1} and π' differ. Then $d_j^{(i-1)} < d'_j$ and we construct $\pi_i : d_0^{(i)}, \ldots, d_{n-1}^{(i)}$ by

$$d_j^{(i)} = d_j^{(i-1)} + 1, \quad d_{j+1}^{(i)} = d_{j+1}^{(i-1)} - 1, \quad \text{and} \quad d_l^{(i)} = d_l^{(i-1)} \text{ otherwise.}$$

If necessary, π_i is sorted in nonincreasing order. Then π_i is again a tree sequence and the theorem is proved by iteratively applying the previous argument to consecutive pairs (π_{i-1}, π_i) of degree sequences that differ in only two positions.   □

Liu et al. [93] have provided an example showing that Theorem 3.13 cannot be directly extended to connected graphs. Let $\pi : 4, 3, 3, 3, 2, 2, 1$ and $\pi' : 4, 4, 3, 2, 2, 2, 1$. The graphs G_π and $G_{\pi'}$ shown in Fig. 3.1 have the maximum spectral radius in $\mathcal{C}_\pi$ and $\mathcal{C}_{\pi'}$, respectively, $\pi \lhd \pi'$ holds, yet $3.09787 \approx \lambda_1(G_\pi) > \lambda_1(G_{\pi'}) \approx 3.05401$.

Liu and Liu [92] have later shown that Theorem 3.13 holds for pairs of degree sequences corresponding to unicyclic graphs, and that it also holds for pairs of degree sequences, for which one degree sequence is normally majorized by another degree sequence.

Definition 3.2 ([92]). Let $\pi : d_0 \geq d_1 \geq \cdots \geq d_{n-1}$ and $\pi' : d'_0 \geq d'_1 \geq \cdots \geq d'_{n-1}$ be two degree sequences such that $\sum_{i=0}^{n-1} d_i = \sum_{i=0}^{n-1}$

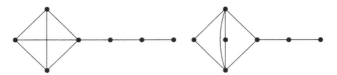

Figure 3.1 Graphs G_π and $G_{\pi'}$ [93].

$d'_i = 2(n - 1 + k)$. If $\pi \lhd \pi'$ and there exists t, $0 \leq t \leq n - 1$, such that $d'_t \geq k+1$ and $d_i = d'_i$ for all $t+1 \leq i \leq n-1$, then π is normally majorized by π'.

Theorem 3.14 ([92]). *Let π and π' be two distinct degree sequences, such that π is normally majorized by π'. If G and G' are the graphs with the maximum spectral radius in C_π and $C_{\pi'}$, respectively, then $\lambda_1(G) < \lambda_1(G')$.*

3.3 GRAPHS WITH A FEW EDGES

Yuan Hong [82] has shown that for a graph with m edges holds

$$\lambda_1 \leq \sqrt{2m - n + 1}. \tag{3.7}$$

This shows that in classes of graphs with linear number of edges, such as trees or planar graphs, the spectral radius is of order $O(\sqrt{n})$. However, if we restrict the maximum vertex degree to Δ in those classes, one can show that λ_1 is actually of order $O(\sqrt{\Delta})$, regardless of the number of vertices n, as we will see in the next two subsections.

3.3.1 Trees

If a tree T has the maximum vertex degree Δ, then

$$\lambda_1 < 2\sqrt{\Delta - 1}. \tag{3.8}$$

This bound has a few different proofs, and we will present two of them.

Firstly, (3.8) follows from the facts that any finite tree with the maximum vertex degree Δ is a proper subgraph of the infinite, Δ-regular tree and that $2\sqrt{\Delta - 1}$ is the spectral radius of the infinite, Δ-regular tree. This is a folklore result, but as it is not straightforward to track it down nowadays, we repeat here its proof from [64].

Theorem 3.15 ([64]). *The spectrum of the infinite, Δ-regular tree is the interval $[-2\sqrt{\Delta - 1}, 2\sqrt{\Delta - 1}]$.*

Proof. Fix v to be an arbitrary vertex of the infinite, Δ-regular tree T with the vertex set V and the adjacency operator A. Consider the *radial* function $f_r \colon V \to \mathbb{C}$ whose value at the ith level of T as rooted at v, i.e., at all vertices at distance i from v, is r^i. We have that for $w \in V$

$$((A - \lambda I)f_r)(w) = \begin{cases} \Delta r - \lambda, & \text{if } w = v, \\ r^{i-1}\left((\Delta - 1)r^2 - \lambda r + 1\right) & \text{if } w \text{ lies on level } i \geq 1. \end{cases}$$

Fix a real λ with $|\lambda| > 2\sqrt{\Delta - 1}$. There exists a solution r to the equation

$$(\Delta - 1)r^2 - \lambda r + 1 = 0 \qquad (3.9)$$

with $|r| < (\Delta - 1)^{-1/2}$, which makes the resulting f_r lie in $l^2(V)$. For such r we have $\Delta r - \lambda \neq 0$, and therefore the equation in x

$$(A - \lambda I)x = \delta_v \qquad (3.10)$$

has a solution $x_v \in l^2(V)$, where δ_v is the Kronecker's delta function that is equal to 1 on v and 0 elsewhere. Writing an arbitrary $y \in l^2(V)$ as a (possibly infinite) linear combination of such δ's and using linearity, we can solve the above equation with δ_v replaced by any y, with $||x||$ bounded by a constant times $||w||$. Therefore, $A - \lambda I$ is invertible and λ does not belong to the spectrum of A.

On the other hand, if $|\lambda| < 2\sqrt{\Delta - 1}$, (3.10) has no solution in $l^2(V)$. Indeed, if such x would exist, then its symmetrization $\tilde{x}$, whose value at each vertex on the ith level is the average of the ith level value of x, would also be an $l^2(V)$ solution to (3.10). Then the values $\tilde{x}_i$ of x at the ith level would satisfy

$$(\Delta - 1)\tilde{x}_{i+2} - \lambda\tilde{x}_{i+1} + \tilde{x}_i = 0, \quad \forall i \geq 0,$$

and so,

$$\tilde{x}_i = c_1 r_1^i + c_2 r_2^i$$

for some constants c_1, c_2 and r_1, r_2 being the roots of (3.9). But for $|\lambda| < 2\sqrt{\Delta - 1}$, both roots r_1, r_2 are of the modulus $(\Delta - 1)^{-1/2}$, contradicting the fact that $\tilde{x} \in l^2(V)$.

Finally, $\lambda = \pm 2\sqrt{\Delta - 1}$ is in the continuous spectrum of A, as the spectrum is a closed set. Since A is self-adjoint, its spectrum consists of real numbers, and we have fully determined it.   □

Back to the setting of finite graphs, the bound (3.8) has been proved by Godsil in [68] by resorting to the matching theory and a result on partition functions from [77, Theorem 4.3]. It has been later proved by Stevanović [142] by relying on the principal eigenvector of particular subtrees of the infinite, Δ-regular tree.

Definition 3.3 ([77]). The rooted Bethe tree $B_{\Delta,1}$ is a single vertex, which is simultaneously the root. For $k > 1$, the tree $B_{\Delta,k}$ consists of the root vertex u which is joined by edges to the roots of each of $\Delta - 1$ copies of $B_{\Delta,k-1}$.

Observe that any tree T with the maximum degree Δ is an (induced) subgraph of $B_{\Delta,D}$, where D is the diameter of T, as it is enough to take any vertex u of T with degree less than Δ and to map it to the root of $B_{\Delta,D}$, with vertices of T at distance i from u being mapped to the vertices on the ith layer of $B_{\Delta,D}$. The bound (3.8) now follows from the monotonicity of the spectral radius and the following result.

Theorem 3.16 ([142]). $\lambda_1(B_{\Delta,k}) < 2\sqrt{\Delta - 1}$ *for each $k \in \mathbb{N}$.*

Proof. Let λ_1 and x be the spectral radius and the principal eigenvector of $B_{\Delta,k}$. As the vertices of the ith level of $B_{\Delta,k}$ are similar to each other, we may denote by x_i the common principal eigenvector component of vertices of the ith level of $B_{\Delta,k}$. Further, let b_i be the number of vertices of the ith level of $B_{\Delta,k}$:

$$b_0 = 1,$$
$$b_i = (\Delta - 1)b_{i-1}, \quad i = 1, \dots, k,$$

so that $b_i = (\Delta - 1)^i$, $i = 0, \dots, k$. The condition $||x|| = 1$ now implies

$$\sum_{i=0}^{k} b_i x_i^2 = 1.$$

The Rayleigh quotient $\lambda_1 = 2\sum_{uv \in E(B_{\Delta,k})} x_u x_v$ can be rewritten as

$$\lambda_1 = 2\sum_{i=1}^{k} b_i x_{i-1} x_i.$$

Now, the Cauchy-Schwarz inequality applied to the vectors

$$p = (\sqrt{b_1}x_0, \sqrt{b_2}x_1, \dots, \sqrt{b_k}x_{k-1}) \quad \text{and} \quad q = (\sqrt{b_1}x_1, \sqrt{b_2}x_2, \dots, \sqrt{b_k}x_k)$$

yields

$$\lambda_1 = 2p \cdot q$$
$$\leq 2\sqrt{b_1 x_0^2 + b_2 x_1^2 \cdots + b_k x_{k-1}^2}\sqrt{b_1 x_1^2 + b_2 x_2^2 + \cdots + b_k x_k^2}$$
$$= 2\sqrt{(\Delta - 1)(b_0 x_0^2 + b_1 x_1^2 + \dots + b_{k-1} x_{k-1}^2)}\sqrt{b_1 x_1^2 + b_2 x_2^2 + \dots + b_k x_k^2}$$

$$= 2\sqrt{(\Delta - 1)(1 - b_k x_k^2)}\sqrt{1 - x_0^2}$$
$$< 2\sqrt{\Delta - 1}.$$

$\square$

Rojo and Robbiano [125] have later shown that

$$\lambda_1(B_{\Delta,k}) = 2\sqrt{\Delta - 1}\cos\frac{\pi}{k+1}. \tag{3.11}$$

They have also determined the full spectrum of $B_{\Delta,k}$, as well as that of the generalized Bethe tree $BT_{d_0,d_1,...,d_{h-1}}$, in which all vertices at the level i have degree d_i (which are not necessarily equal), with vertices at the level h all having degree 1. In addition, Song et al. [138] have observed that $\lambda_1(BT_{\Delta,\Delta}) = \sqrt{2\Delta - 1}$, so that if $n \leq \Delta^2 + 1$ holds for a tree T, then $\lambda_1(T) \leq \sqrt{2\Delta - 1}$.

Simić and Tošić [136] have described the structure of the tree with the maximum spectral radius among trees with n vertices and the maximum vertex degree Δ. They have shown that such a tree is formed from $BT_{\Delta,...,\Delta}$ by attaching $\Delta - 1$ new edges to its leaves, consecutively from left to right, and attaching the remaining number of edges (if nonzero) to the next leaf. A particular case of this result for trees with the maximum vertex degree 4 has been proved independently by Biyikoğlu and Leydold [18] by resorting to Theorem 3.13.

On the opposite side, Belardo et al. [9] have shown that the tree with the minimum spectral radius in the set of trees whose n vertices have degrees either Δ or 1, is necessarily a (unique) caterpillar in that set.

Hu [84] have shown that the similar bound as in (3.8) holds for unicyclic graphs as well, by providing an appropriate orientation of its edges.

Theorem 3.17 ([84]). *If G is a unicyclic graph with the maximum vertex degree Δ, then*

$$\lambda_1 \leq 2\sqrt{\Delta - 1},$$

with equality if and only if $G \cong C_n$.

Proof. Let λ_1 and x be the spectral radius and the principal eigenvector of G. From the Rayleigh quotient, we know that

$$\lambda_1 = 2 \sum_{ij \in E(G)} x_i x_j.$$

Let C_k denote the unique cycle in G. Denote the vertices of C_k by $1, \ldots, k$ and orient the edges of C_k as $i \rightarrow i+1$, $i = 1, \ldots, k$ (with $k + 1 \equiv 1$ (mod k)). The graph $G - C_k$ consists of k trees. Denote the vertices of $G - C_k$ by $k + 1, \ldots, n$ and orient the edges of $G - C_k$ in such a way that there exists a directed path from each pendant vertex of G to the closest vertex on C_k.

For each vertex v of G, denote by p_v the tail of the directed edge having v as its head under the above orientation. Since $|E(G)| = |V(G)|$ for unicyclic graphs, we have

$$\lambda_1 = 2 \sum_{ij \in E(G)} x_i x_j = 2 \sum_{v \in V(G)} x_v x_{p_v}. \tag{3.12}$$

Under the given orientation, the outdegree of each vertex v is 1, while its indegree is $\deg_v - 1$. Thus,

$$\sum_{v \in V(G)} x_{p_v}^2 = \sum_{v \in V(G)} (\deg_v - 1) x_v^2.$$

The Cauchy-Schwarz inequality applied to (3.12) now yields

$$\begin{aligned}
\lambda_1 &= 2 \sum_{v \in V(G)} x_v x_{p_v} \\
&\leq 2 \sqrt{\sum_{v \in V(G)} x_v^2} \sqrt{\sum_{v \in V(G)} x_{p_v}^2} \\
&= 2\sqrt{1} \sqrt{\sum_{v \in V(G)} (\deg_v - 1) x_v^2} \\
&\leq 2 \sqrt{(\Delta - 1) \sum_{v \in V(G)} x_v^2} \\
&= 2\sqrt{\Delta - 1}.
\end{aligned}$$

If the equality holds above, then $\sum_{v \in V(G)} (\deg_v - 1) x_v^2 = \Delta - 1$, showing that $\deg_v = \Delta$ for each $v \in V(G)$, which is possible if and only if $\Delta = 2$ and G is the cycle C_n. On the other hand, the equality does hold for $G \cong C_n$. $\square$

3.3.2 Planar Graphs

Schwenk and Wilson [132] asked in 1978 what can be said about the eigenvalues of a planar graph. From the Hong's bound (3.7) and the fact that a planar graph has at most $3n - 6$ edges, one immediately gets that for a planar graph [82]

$$\lambda_1 \leq \sqrt{5n - 11}.$$

This bound can, however, be improved by exploiting the Weyl inequalities (see Theorem 1.8) and edge decompositions of planar graphs, as shown first by Cao and Vince [26].

Theorem 3.18 ([26]). *If G is a planar graph with $n \geq 3$ vertices, then*

$$\lambda_1 \leq \sqrt{3n - 9} + 4.$$

Cao and Vince relied on the fact that the maximal planar graph, containing G as a spanning subgraph, may be decomposed into a spanning tree T with the maximum vertex degree four and a spanning subgraph H with no isolated vertices. For tree T holds

$$\lambda_1(T) < 2\sqrt{3} < 4$$

by (3.8). The subgraph H has at most $(3n - 6) - (n - 1) = 2n - 5$ edges, so that by Hong's bound (3.7)

$$\lambda_1(H) \leq \sqrt{2(2n - 5) - n + 1} = \sqrt{3n - 9}.$$

Finally, from the Weyl inequality $\lambda_1(G) \leq \lambda_1(T) + \lambda_1(H) < \sqrt{3n - 9} + 2\sqrt{3}$.

Cao and Vince also posed the conjecture on the planar graph with the maximum spectral radius.

Conjecture 3.2 ([26]). *If G is a planar graph with n vertices, then*

$$\lambda_1(G) \leq \lambda_1(K_2 \vee P_{n-2}).$$

Interestingly, this conjecture was also posed by Boots and Royle [20] two years earlier in a geographical journal, where the spectral radius was applied in the study of geographical networks.

The following result of Hayes [75] is useful if planar graphs are restricted to the fixed maximum vertex degree Δ.

Theorem 3.19 ([75]). *Let G be a simple graph with the maximum vertex degree Δ, whose edges can be oriented such that the maximum outdegree d is at most $\Delta/2$. Then*

$$\lambda_1 \leq 2\sqrt{d(\Delta - d)}.$$

Proof. From Theorem 1.1 the number W_l of all walks of length l in G is equal to the sum of the entries of A^l, for $l \geq 0$. Let $\lambda_1 \geq \cdots \geq \lambda_n$ be the eigenvalues of A, with corresponding orthonormal eigenvectors $x_1, \ldots, x_n$. Let j be the all-one vector and let $j = \sum_{i=1}^{n} \alpha_i x_i$ be the representation of j in the basis $x_1, \ldots, x_n$. From the spectral decomposition (1.1) of A, it follows that

$$W_l = j^T A j = \sum_{i=1}^{n} \lambda_i^l (j^T x_i)(x_i^T j) = \sum_{i=1}^{n} \lambda_i^l \alpha_i^2 = \lambda_1^l \sum_{i=1}^{n} \alpha_i^2 \left(\frac{\lambda_i}{\lambda_1} \right)^l.$$

Here $\alpha_1 = x_1^T j > 0$, due to the positivity of x_1. If G is not bipartite, then $|\lambda_i| < \lambda_1$ for $i = 2, \ldots, n$, so that

$$\lim_{l \to \infty} (W_l)^{1/l} = \lambda_1.$$

If G is bipartite, then $\lambda_{n+1-i} = -\lambda_i$ for $i = 1, \ldots, n$, and $W_l = 0$ for odd l. For even $l = 2k$, however, we still have

$$\lim_{k \to \infty} (W_{2k})^{1/(2k)} = \lambda_1.$$

In light of this, in order to show that $\lambda_1 \leq 2\sqrt{d(\Delta - d)}$ it is enough to show that

$$W_{2k} \leq 2n \, 2^{2k} d^k (\Delta - d)^k. \tag{3.13}$$

(The constant factor $2n$ above is irrelevant for the limit as $\lim_{k \to \infty} (2n)^{1/(2k)} = 1$.)

The facts that the maximum vertex degree is Δ and that the edges of G can be oriented in such a way that the outdegree of each vertex is at most d, mean that for each vertex v of G, its neighbors can be labeled with distinct labels from $\{1, \ldots, \Delta\}$, so that the out-neighbors of v are given the labels between 1 and d. (it will not make a problem if some of the in-neighbors of v are also given a label at most d). Note, in addition, that during this labeling process each vertex of G gets a separate label from each of its neighbors.

Now, let $S: v_0, \ldots, v_{2k}$ denote an arbitrary walk of length $2k$ in G, and let F_S denote the set of steps i such that v_{i+1} has a label at most d as a neighbor of v_i. The crucial observation here is that at least k of the edges $v_0 v_1, v_1 v_2, \ldots, v_{2k-1} v_{2k}$ of S have the same orientation, so that

$$|F_S| \geq k \quad \text{or} \quad |F_{R(S)}| \geq k,$$

where $R(S): v_{2k}, \ldots, v_0$ denotes the reverse walk of S.

We can now bound the number of all walks W_{2k} from above by firstly counting all walks S with $|F_S| \geq k$, and then by multiplying the resulting amount with 2, in order to account for those walks with $|F_S| < k$ and $|F_{R(S)}| \geq k$.

For a fixed subset $F \subseteq \{0, \ldots, 2k - 1\}$, there exists at most

$$n d^{|F|} (\Delta - d)^{2k - |F|}$$

walks S such that $F_S = F$. The factor n denotes the starting vertex of a walk, while $d^{|F|}$ and $(\Delta - d)^{2k-|F|}$ denote the numbers of choices for the next vertex in the walk, according to F. Hence, the number of walks S with $|F_S| \geq k$ is at most

$$\sum_{|F|=k}^{2k} n d^{|F|} (\Delta - d)^{2k-|F|} \leq \sum_{|F|=k}^{2k} n d^k (\Delta - d)^k$$

due to $d \leq \Delta - d$. The number of subsets $F \subseteq \{0, \ldots, 2k-1\}$ with $k \leq |F| \leq 2k$ is certainly less than 2^{2k}, the number of all subsets of $\{0, \ldots, 2k - 1\}$, so that the number of walks S with $|F_S| \geq k$ is at most $n 2^{2k} d^k (\Delta - d)^k$. From the above observation, we then have

$$W_{2k} \leq 2n \, 2^{2k} d^k (\Delta - d)^k,$$

which establishes (3.13). $\qquad\qquad\qquad\qquad\qquad\qquad\qquad\qquad \square$

By reversing orientation of edges, the previous theorem also holds if we replace outdegree with indegree. Combined with the result of Chrobak and Eppstein [30] that every planar graph admits an orientation with maximum indegree three, we obtain

Corollary 3.1 ([75]). *If G is a planar graph with the maximum vertex degree* $\Delta \geq 6$, *then*

$$\lambda_1 \leq \sqrt{12(\Delta - 3)}.$$

Dvořák and Mohar [54] improved previous corollary by relying on the following decomposition result of Gonçalves [70].

Theorem 3.20 ([70]). *If G is a planar graph, then* $E(G) = E(T_1) \cup E(T_2) \cup E(T_3)$, *where* T_1, T_2, *and* T_3 *are forests and* $\Delta(T_3) \leq 4$.

Theorem 3.21 ([54]). *If G is a planar graph with the maximum vertex degree* $\Delta \geq 2$, *then*

$$\lambda_1 \leq \sqrt{8(\Delta - 2)} + 2\sqrt{3}. \tag{3.14}$$

Proof. If $\Delta \leq 3$, then the claim follows as $\lambda_1 \leq \Delta$. Assume, therefore, that $\Delta \geq 4$. Let $E(G) = E(T_1) \cup E(T_2) \cup E(T_3)$ be the forest decomposition of G from Theorem 3.20. Then T_1 and T_2 admit an orientation with the maximum indegree 1; thus $T_1 \cup T_2$ has an orientation with the maximum indegree at most 2. Then from Theorem 3.19 we have

$$\lambda_1(T_1 \cup T_2) \leq 2\sqrt{2(\Delta - 2)}. \tag{3.15}$$

Since T_3 is a tree with the maximum vertex degree at most 4, we have

$$\lambda_1(T_3) < 2\sqrt{3}. \tag{3.16}$$

Applying the Weyl inequality to (3.15) and (3.16) then yields (3.14).   □

Dvořák and Mohar [54] have further improved (3.14) for planar graphs with higher connectivity and provided analogous results for graphs embedded in surface of given Euler genus.

Hayes' Theorem 3.19 can be further applied to d-degenerate graphs.

Definition 3.4. A graph G is d-degenerate if every subgraph of G has a vertex of degree at most d.

The above condition is equivalent to the requirement that G can be reduced to the empty graph by successively removing vertices whose degree is at most d. If at each step of this removal process we orient the edges of G

from the removed vertex to its neighbors, then we see that the d-degenerate graph G admits an orientation with maximum outdegree d. Theorem 3.19 then shows that the spectral radius of G with $\Delta \geq 2d$ is at most $2\sqrt{d(\Delta - d)}$. Moreover, this holds for every subgraph H of G, as H is also d-degenerate, suggesting the following

Definition 3.5 ([55]). A graph G is spectrally d-degenerate if for every subgraph H of G holds $\lambda_1(H) \leq \sqrt{d\Delta_H}$, where Δ_H is the maximum vertex degree of H.

Thus, every d-degenerate graph is spectrally $4d$-degenerate by Theorem 3.19. Dvořák and Mohar [55] proved a rough converse by showing that a spectrally d-degenerate graph contains a vertex whose degree is at most $\max\{4d, 4d\log_2(\Delta/d)\}$. However, Alon [1] showed that the full converse cannot hold, by providing a probabilistic construction for the following

Theorem 3.22 ([1]). *For every positive integer M, there exists a spectrally 50-degenerate graph G in which every degree is at least M.*

3.4 COMPLETE MULTIPARTITE GRAPHS

A spectral characterization of complete multipartite graphs is given in the following, old result of Petrović [122], that was originally stated in terms of infinite graphs.

Theorem 3.23 ([122]). *A connected graph has exactly one positive eigenvalue of its adjacency matrix if and only if it is a complete multipartite graph.*

One direction of the above theorem follows from the description of the spectrum of complete multipartite graphs.

Theorem 3.24 ([122]). *For a complete multipartite graph $K_{n_1,\ldots,n_p}$ with $p \geq 2$, let $n'_1 < \cdots < n'_{p'}$ be the distinct values among $n_1,\ldots,n_p$, and let s_i, $i = 1,\ldots,p'$, be the multiplicity of n'_i among $n_1,\ldots,n_p$. The adjacency spectrum of $K_{n_1,n_2,\ldots,n_p}$ has*

1) an eigenvalue 0 of multiplicity $n - p$,

2) *an eigenvalue* $-n_i'$, $i = 1, \ldots, p'$ *of multiplicity* $s_i - 1$, *whenever* $s_i \geq 2$, *and*

3) *exactly* p' *eigenvalues* λ, *distinct from* 0 *and* $-n_i'$, $i = 1, \ldots, p'$, *satisfying*

$$\sum_{i=1}^{p} \frac{n_i}{\lambda + n_i} = 1. \tag{3.17}$$

The above description of the spectrum is obtained easily from the characterization of the eigenvectors of $K_{n_1,\ldots,n_p}$. Let $V = V(K_{n_1,\ldots,n_p})$. Denote the parts of V by $V_1, \ldots, V_p$, such that $n_i = |V_i|$, $i = 1, \ldots, p$, and denote further the vertices of V_i by $v_{i,1}, \ldots, v_{i,n_i}$. For each $i = 1, \ldots, p$, if $n_i \geq 2$, let $x_{i,q} \in R^V$, $q = 2, \ldots, n_i$, denote the vector with

$$x_{i,q}(v_{i,1}) = 1, \quad x_{i,q}(v_{i,q}) = -1, \quad x_{i,q}(v) = 0 \text{ for } v \in V \setminus \{v_{i,1}, v_{i,q}\}.$$

Let $\mathrm{col}_i(A)$ denote the ith column vector of A, the adjacency matrix of $K_{n_1,\ldots,n_p}$. Then

$$A x_{i,q} = \mathrm{col}_{v_{i,1}}(A) - \mathrm{col}_{v_{i,q}}(A) = 0,$$

as $v_{i,1}$ and $v_{i,q}$ have equal neighborhoods. Hence $x_{i,q}$, $i = 1, \ldots, p$, $q = 2, \ldots, n_i$, are linearly independent eigenvectors of the eigenvalue 0 of A, showing that the multiplicity of 0 is at least $\sum_{i=1}^{p}(n_i - 1) = n - p$. The remaining p eigenvectors can be chosen so that they are orthogonal to all of $x_{i,q}$. Such an eigenvector y has to be constant within each part V_i, $i = 1, \ldots, p$. If $n_i \geq 2$, then the orthogonality of y and $x_{i,q}$ implies that $y(v_{i,1}) = y(v_{i,q})$ for each $q = 2, \ldots, n_i$ (and if $n_i = 1$, then y is trivially constant on the singleton V_i).

Let $j_i \in R^V$, $i = 1, \ldots, p$, be the vector equal to 1 for each vertex of V_i, and equal to 0 for vertices in $V \setminus V_i$. Then there exist coefficients $\alpha_i \in R$, $i = 1, \ldots, p$, such that

$$y = \sum_{i=1}^{p} \alpha_i j_i.$$

Let λ be an eigenvalue of A corresponding to the eigenvector y. Noting that the column of A corresponding to a vertex in V_i is equal to $j - j_i$, $i = 1, \ldots, p$, we get

$$0 = Ay - \lambda y = \sum_{i=1}^{p} \alpha_i(Aj_i - \lambda j_i)$$

$$= \sum_{i=1}^{p} \alpha_i (n_i(j - j_i) - \lambda j_i)$$

$$= \sum_{i=1}^{p} \alpha_i n_i \sum_{k \neq i} j_k - \sum_{i=1}^{p} \alpha_i \lambda j_i$$

$$= \sum_{i=1}^{p} j_i \left(\sum_{k \neq i} \alpha_k n_k - \alpha_i \lambda \right).$$

Let $c = \sum_{k=1}^{p} \alpha_k n_k$. Then

$$0 = \sum_{i=1}^{p} j_i \left(c - \alpha_i(n_i + \lambda) \right). \tag{3.18}$$

The vectors j_i, $i = 1, \ldots, p$, are linearly independent, implying that the coefficient of each j_i in (3.18) is 0. Hence for each $i = 1, \ldots, p$,

$$\alpha_i(n_i + \lambda) = \sum_{k=1}^{p} \alpha_k n_k. \tag{3.19}$$

If $\lambda = -n_i$ for some $i = 1, \ldots, p$, then $c = 0$ and, consequently, $\alpha_k = 0$ hold for each k such that $n_k \neq n_i$. Let K be the set of all indices k such that $n_k = n_i$. Then (3.19) reduces to

$$\sum_{k \in K} \alpha_k = 0,$$

which has $|K| - 1$ linearly independent solutions, yielding that $-n_i$ is an eigenvalue with multiplicity $|K| - 1$.

If $\lambda + n_i \neq 0$ for each i, then from (3.19) we first get that for each $i = 1, \ldots, p$,

$$\alpha_i = \frac{c}{n_i + \lambda}, \tag{3.20}$$

and then, replacing α_i back into $c = \sum_{k=1}^{p} \alpha_k n_k$ and dividing by c, we obtain

$$f(\lambda) := \sum_{k=1}^{p} \frac{n_k}{n_k + \lambda} = 1. \tag{3.21}$$

Recall that $n'_1 < \cdots < n'_{p'}$ are the distinct values among $n_1, \ldots, n_p$. The function f is defined on the union of intervals $(-\infty, -n'_{p'}) \cup (-n'_{p'}, -n'_{p'-1}) \cup \cdots \cup (-n'_2, -n'_1) \cup (-n'_1, +\infty)$. The derivative $f'(\lambda) = \sum_{k=1}^{p} -\frac{n_k}{(n_k+\lambda)^2}$ is negative everywhere, so that $f(\lambda)$ is strictly decreasing on each interval. The function $f(\lambda)$ is negative on $(-\infty, -n'_{p'})$, so that (3.21) has no solution on this interval. Since

$$\lim_{\lambda \to -n'_i-} f(\lambda) = -\infty,$$

$$\lim_{\lambda \to -n'_i+} f(\lambda) = +\infty,$$

$$\lim_{\lambda \to +\infty} f(\lambda) = 0,$$

(3.21) has a unique solution on each of the intervals $(-n'_i, -n'_{i-1})$, $i = 2, \ldots, p'$, and $(-n'_1, +\infty)$. Since $f(0) = p \geq 2$, the unique solution in $(-n'_1, +\infty)$ actually belongs to $(0, +\infty)$, yielding that $K_{n_1,\ldots,n_p}$ has exactly one positive eigenvalue.

For the proof of the other direction of Theorem 3.23, Petrović [122] relied on a simple and instructive use of the Interlacing theorem (see Theorem 1.6): if a connected graph is not a complete multipartite graph, then it contains two nonadjacent vertices a and b, together with a vertex c that is, say, adjacent to b, but not to a. Since a is not an isolated vertex, there exists a vertex d adjacent to a (see Fig. 3.2). Then, regardless of the existence of edges bd and cd, the subgraph induced by the vertices a, b, c, and d has two positive eigenvalues, implying that a graph which is not a complete multipartite graph has at least two positive eigenvalues.

The spectral radius of the complete multipartite graph $K_{n_1,\ldots,n_p}$ can be easily determined from (3.17) if it has at most two different values among $n_1, \ldots, n_p$. So, for example,

$$\lambda_1(K_{n_1}) = \lambda_1(K_{1,1,\ldots,1}) = n_1 - 1 \quad \text{and} \quad \lambda_1(K_{n_1,n_2}) = \sqrt{n_1 n_2}.$$

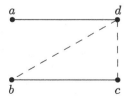

Figure 3.2 The subgraph induced by the vertices a, b, c, and d.

Stevanović et al. [145] have recently characterized complete multipartite graphs with the extremal values of the spectral radius. The maximum part of the following theorem has been obtained earlier by Feng et al. [59] as well.

Theorem 3.25 ([59, 145]). *Let n and k be fixed and $n \geq k \geq 2$. Among complete multipartite graphs $K_{n_1,\ldots,n_p}$ with $n_1 + \cdots + n_p = n$, the minimum spectral radius is attained by the complete split graph $CS_{n,p-1}$, while the maximum spectral radius is attained by Turán graph $T_{n,p}$.*

Proof. Let λ_1, x, and E be, respectively, the spectral radius, the principal eigenvector, and the edge set of $K_{n_1,\ldots,n_p}$. Since λ_1 is a simple eigenvalue of $K_{n_1,\ldots,n_p}$, similar vertices have equal principal eigenvector components. Hence, we may denote by x_i the common x-component of vertices in the part of $K_{n_1,\ldots,n_p}$ having n_i vertices for $i = 1,\ldots,p$. From the eigenvalue equation we have

$$\lambda_1 x_i = \sum_{\substack{k=1 \\ k \neq i}}^{p} n_k x_k = X - n_i x_i,$$

where $X = \sum_{k=1}^{p} n_k x_k$. Then

$$x_i = \frac{X}{\lambda_1 + n_i}. \tag{3.22}$$

Next, we show that if $n_i - n_j \geq 2$, then

$$\lambda_1(K_{n_1,\ldots,n_i-1,\ldots,n_j+1,\ldots,n_p}) > \lambda_1(K_{n_1,\ldots,n_i,\ldots,n_j,\ldots,n_p}). \tag{3.23}$$

Let λ_1^*, A^*, and E^* denote, respectively, the spectral radius, the adjacency matrix, and the edge set of $K_{n_1,\ldots,n_i-1,\ldots,n_j+1,\ldots,n_p}$. From the Rayleigh quotient and (3.22) we have

$$\begin{aligned}
\lambda_1^* \geq x^T A^* x &= \sum_{uv \in E^*} 2x_u x_v \\
&= \sum_{uv \in E} 2x_u x_v + \sum_{uv \in E^* \setminus E} 2x_u x_v - \sum_{uv \in E \setminus E^*} 2x_u x_v \\
&= \lambda_1 + 2x_i(n_i - 1)x_i - 2x_i n_j x_j \\
&= \lambda_1 + 2x_i X \left(\frac{n_i - 1}{\lambda_1 + n_i} - \frac{n_j}{\lambda_1 + n_j} \right). \tag{3.24}
\end{aligned}$$

Next, note that $K_{n_1,\ldots,n_i,\ldots,n_j,\ldots,n_p}$ has K_{n_i,n_j} as an induced subgraph, so that, by the Interlacing theorem

$$\lambda_1 \geq \sqrt{n_i\, n_j} > n_j.$$

Therefore,

$$\frac{n_i - 1}{\lambda_1 + n_i} - \frac{n_j}{\lambda_1 + n_j} = \frac{(n_i - n_j - 1)\lambda_1 - n_j}{(\lambda_1 + n_i)(\lambda_1 + n_j)} \geq \frac{\lambda_1 - n_j}{(\lambda_1 + n_i)(\lambda_1 + n_j)} > 0,$$

so that $\lambda_1^* > \lambda_1$ follows from (3.24).

Now, let $K_{m_1,\ldots,m_p}$ be the complete multipartite graph with the minimum spectral radius. If there are two parameters m_i and m_j, such that $m_i \geq m_j \geq 2$, then $(m_i + 1) - (m_j - 1) \geq 2$ and from (3.23)

$$\lambda_1(K_{m_1,\ldots,m_i,\ldots,m_j,\ldots,m_p}) > \lambda_1(K_{m_1,\ldots,m_i+1,\ldots,m_j-1,\ldots,m_p})$$

contradicting the choice of $K_{m_1,\ldots,m_i,\ldots,m_j,\ldots,m_p}$. Hence, all parameters $m_1,\ldots,m_p$ must be equal to 1, except for one parameter equal to $n - p + 1$, so that $K_{m_1,\ldots,m_p} \cong CS_{n,p-1}$ has the minimum spectral radius.

Next, let $K_{m_1,\ldots,m_p}$ be the complete multipartite graph with the maximum spectral radius. It is apparent from (3.23) that $|m_i - m_j| \leq 1$ holds for all $i \neq j$, as otherwise, assuming $m_i - m_j \geq 2$,

$$\lambda_1(K_{m_1,\ldots,m_i-1,\ldots,m_j+1,\ldots,m_p}) > \lambda_1(K_{m_1,\ldots,m_i,\ldots,m_j,\ldots,m_p})$$

contradicting the choice of $K_{m_1,\ldots,m_i,\ldots,m_j,\ldots,m_p}$. The condition $|m_i - m_j| \leq 1$ for $i \neq j$ implies that each parameter m_i is equal to either $\lfloor n/p \rfloor$ or $\lceil n/p \rceil$, so that $K_{m_1,\ldots,m_p} \cong T_{n,p}$ has the maximum spectral radius. $\square$

Previous theorem could have been proved also by Delorme's observation [49] that the spectral radius of a complete multipartite graph increases by replacing arbitrary r parameters of a complete multipartite graph with their average value. Delorme proved this by considering the change in the coefficients of the characteristic polynomial of the complete multipartite graph, which are given by the following

Theorem 3.26 ([49]). *The characteristic polynomial of the complete multipartite graph $K_{n_1,\ldots,n_p}$ is*

$$P_{K_{n_1,\ldots,n_p}}(\lambda) = \lambda^{n-p}\left(\lambda^p - \sum_{m=2}^{p}(m-1)\sigma_m(n_1,\ldots,n_p)\lambda^{p-m}\right), \qquad (3.25)$$

where

$$\sigma_m(n_1, \ldots, n_p) = \sum_{I \subseteq \{1, \ldots, k\}, |I| = m} \prod_{i \in I} n_i$$

is the elementary symmetric function of degree m.

Note that the above formula, although without proof, appeared earlier in [73], where it was attributed to M.D. Plummer. Another formula for the characteristic polynomial of the complete multipartite graph [43, p. 74] has the form

$$P_{K_{n_1, \ldots, n_p}}(\lambda) = \lambda^{n-p} \left(1 - \sum_{i=1}^{p} \frac{n_i}{\lambda + n_i} \right) \prod_{j=1}^{p} (\lambda + n_j). \qquad (3.26)$$

In particular, Feng et al. [59] have used (3.26) to prove (3.23) in another way.

Delorme [49] has also asked whether the spectral radius of $K_{n_1, \ldots, n_k}$ is a *concave* function of $(n_1, \ldots, n_k)$ on the $(k-1)$-dimensional simplex X: $\sum_{i=1}^{k} n_i = n$, $n_i \geq 0$ for $i = 1, \ldots, k$, i.e., whether

$$\lambda_1(K_{t(n_1, \ldots, n_k) + (1-t)(m_1, \ldots, m_k)}) \geq t\lambda_1(K_{n_1, \ldots, n_k}) + (1-t)\lambda_1(K_{m_1, \ldots, m_k})$$

for any two points $(n_1, \ldots, n_k), (m_1, \ldots, m_k) \in X$ and each $t \in [0, 1]$ such that

$$t(n_1, \ldots, n_k) + (1-t)(m_1, \ldots, m_k) \in X.$$

This is clear for $k = 2$ and Delorme has proved it for $k = 3$, although the details have not been given in [49]. Computationally, the question has been affirmatively tested for $t = \frac{1}{2}$, $k \in \{4, \ldots, 10\}$ and $n \leq 33$ in [145].

Similar result to that of Theorem 3.23, but aimed to modularity matrix, has been shown recently by Majstorović and Stevanović [100]. The modularity matrix M of a graph G is defined as

$$M = A - \frac{1}{2m} dd^T, \qquad (3.27)$$

where A, m, and d are, respectively, the adjacency matrix, the number of edges, and the vector of vertex degrees of G. See [108, Chapter 11] for further details and references on modularity matrix. Mastorović and Stevanović state

Theorem 3.27 ([100]). *The largest eigenvalue of the modularity matrix of a connected graph is zero if and only if the graph is a complete multipartite graph.*

In one direction, this theorem is proved by describing the modularity spectrum of complete multipartite graphs, analogous to Theorem 3.24, but for the other direction it uses an interesting relation between the adjacency spectral radius and the largest modularity eigenvalue of graphs. Let us first state a classical lemma.

Lemma 3.4. *Let A be a real symmetric $n \times n$ matrix with eigenvalues $\lambda_1(A) \geq \cdots \geq \lambda_n(A)$. Let $B = A + zz^T$, where $z \in R^n$, have eigenvalues $\lambda_1(B) \geq \cdots \geq \lambda_n(B)$. Then*

$$\lambda_i(B) \geq \lambda_i(A) \geq \lambda_{i+1}(B), \qquad i = 1, \ldots, n,$$

under the convention that $\lambda_{n+1}(B) = -\infty$.

This lemma should have appeared in classical monograph of Gantmacher and Krein [66, pp. 82–86]; the author, however, has no access to it, so that this particular reference is due to [150]. For the sake of completeness, we provide a short proof of the above lemma as adapted from [150]. First, we may assume without loss of generality that A is positive definite, since we may add the same multiple of the unit matrix to both A and B. So, let $A = A_1 A_1^T$ where A_1 is an $n \times n$ matrix. Now the eigenvalues of

$$B = A + zz^T = [A_1 \ z][A_1 \ z]^T$$

are the same, except for an additional zero eigenvalue, as the eigenvalues of the $(n + 1)$-square matrix

$$B' = [A_1 \ z]^T [A_1 \ z] = \begin{bmatrix} A_1^T A_1 & A_1^T z \\ z^T A_1 & z^T z \end{bmatrix}.$$

Since $A_1^T A_1$ is the principal submatrix of B', the eigenvalues $\lambda_1(A) \geq \cdots \geq \lambda_n(A)$ of $A_1^T A_1$ and $\lambda_1(B) \geq \cdots \geq \lambda_n(B) \geq \lambda_{n+1}(B') = 0$ are interlaced:

$$\lambda_i(B) \geq \lambda_i(A) \geq \lambda_{i+1}(B), \qquad i = 1, \ldots, n.$$

Back to the adjacency and modularity matrices, we observe from (3.27) that the modularity matrix M is a rank-one modification of A with $z = d/\sqrt{2m}$, i.e., that

$$A = M + zz^T.$$

Lemma 3.4 now yields that the largest modularity eigenvalue is situated between the two largest adjacency eigenvalues:

$$\lambda_1(A) \geq \lambda_1(M) \geq \lambda_2(A).$$

Hence, the proof of the other direction of Theorem 3.27 stems from Theorem 3.23: if G is not a complete multipartite graph, then it has at least two positive adjacency eigenvalues, implying that its largest modularity eigenvalue is positive.

CHAPTER 4

Spectral Radius and Other Graph Invariants

Graph theorists are relying on computers nowadays to pretest their ideas and conjectures, whether they write their own programs or use ready-made software, such as Graffiti [58], GRAPH [48], newGRAPH [144], AutoGraphiX [27], GrInvIn [121], or MathChem [158]. Notable software among these is AutoGraphiX, which employs the Variable Neighborhood Search metaheuristic in the search for graphs that attain extremal value of user-defined functions in various graph classes. The use of AutoGraphiX has been influential in devising a number of results presented in this chapter, especially in the case of determining extremal values of the spectral radius among connected graphs with a constant value of a selected integer-valued invariant. Some hard conjectures obtained with the help of AutoGraphiX are reviewed in Section 4.1, while results on the graphs with the minimum or the maximum spectral radius among graphs with a given value of the clique number, chromatic number, independence number, matching number, diameter, radius, and domination number are presented in Sections 4.2–4.8. Finally, nearly complete solution to one of the hard conjectures from Section 4.1 is presented in Section 4.9.

4.1 SELECTED AUTOGRAPHIX CONJECTURES

Extremal graph theory deals with the problem of determining extremal values or extremal graphs for a given graph invariant i_G in a given set of graphs $\mathcal{G}$. It has been observed in [27, 28, 44] that this may be viewed as an instance of a parametric combinatorial optimization problem as well, which can be solved with a generic metaheuristic method. This idea has been implemented in the system AutoGraphiX using the Variable Neighborhood Search [103]. Since then, AutoGraphiX has been prominently used to obtain a series of graph theoretical results and conjectures. Those that pertain to graph eigenvalues are extensively surveyed in [4, 5, 7].

We list here some of the tougher conjectures on the spectral radius of graphs obtained with the help of AutoGraphiX, as possible directions for further research.

Spectral Radius of Graphs. http://dx.doi.org/10.1016/B978-0-12-802068-5.00004-X

Conjecture 4.1 ([3, 140]). *The maximum value of $\lambda_1 - \frac{2m}{n}$ among connected graphs with n vertices, $n \geq 10$, is attained by the pineapple $PA_{n,\lceil n/2 \rceil + 1}$.*

As noted in [3, 12], a simple bound on $\lambda_1 - \frac{2m}{n}$, which is close to the true value of this difference for the pineapple $PA_{n,\lceil n/2 \rceil + 1}$, may be obtained by using the Yuan Hong's bound $\lambda_1 \leq \sqrt{2m - n + 1}$ and the geometric-arithmetic mean inequality:

$$
\begin{aligned}
\lambda_1 - \frac{2m}{n} \leq \sqrt{2m - n + 1} - \frac{2m}{n} &= 2\sqrt{\frac{n}{4} \cdot \frac{2m - n + 1}{n}} - \frac{2m}{n} \\
&\leq \frac{n}{4} + \frac{2m - n + 1}{n} - \frac{2m}{n} \\
&= \frac{n}{4} - 1 + \frac{1}{n}.
\end{aligned}
$$

On the other hand, the spectral radius of $PA_{n,\lceil n/2 \rceil + 1}$ is slightly larger than $\lceil \frac{n}{2} \rceil$ (its exact value is a root of a cubic equation, so that the nice closed formula is unavailable), while its average degree is approximately $\frac{n}{4} + \frac{3}{2} - \frac{2}{n}$.

Conjecture 4.1 has been proved for graphs whose minimum vertex degree is at least 2 in [4]. Nevertheless, it is still largely open as even pineapples themselves contain vertices of degree 1.

Note that if we allow disconnected graphs, then the maximum value of $\lambda_1 - \frac{2m}{n}$ has been determined by Bell [12], by relying on Rowlinson's result [126] on the maximum spectral radius of graphs with given numbers of vertices and edges (see the last part of Section 2.4).

Let λ_n denote the smallest eigenvalue of adjacency matrix of a graph. The difference $\lambda_1 - \lambda_n$ is called the spectral spread of a graph. The following conjecture has been independently posed earlier in [71].

Conjecture 4.2 ([3, 140]). *The maximum value of the spectral spread among connected graphs with n vertices is attained by the complete split graph $CS_{n,\lfloor 2n/3 \rfloor}$.*

Conjecture 4.3 ([3, 140]). *The maximum value of $\lambda_1(G) + \lambda_1(\overline{G})$ among simple graphs G on n vertices is attained by the complete split graph $CS_{n,\lceil 2n/3 \rceil}$ and its complement, and if $n \equiv 2 \pmod 3$, also by $CS_{n,\lfloor 2n/3 \rfloor}$ and its complement.*

More precisely, for any n-vertex simple graph G we have

$$\lambda_1(G) + \lambda_1(\overline{G}) \leq \frac{4}{3}n - \frac{5}{3} + \begin{cases} f_1(n) & \text{if } n \equiv 1 \pmod 3, \\ 0 & \text{if } n \equiv 2 \pmod 3, \\ f_2(n) & \text{if } n \equiv 0 \pmod 3, \end{cases}$$

where $f_1(n) = \frac{\sqrt{(3n-2)^2+8}-(3n-2)}{6}$ and $f_2(n) = \frac{\sqrt{(3n-1)^2+8}-(3n-1)}{6}$.

Terpai [149] has come very close to solving this conjecture by proving that

$$\lambda_1(G) + \lambda_1(\overline{G}) \leq \frac{4}{3}n - 1.$$

Instead of considering graphs, he obtained this bound by studying graphons, limit objects of convergent sequences of dense graphs (see the book by Lovász [97] for an introduction to the subject).

Note that AutoGraphiX can be freely downloaded from its web page http://www.gerad.ca/~agx/ so that the reader may easily use it to test own ideas and conjectures.

4.2 CLIQUE NUMBER

The fundamental lemma in the study of spectral properties of graphs with a given clique number is the following result of Motzkin and Straus [107].

Lemma 4.1 ([107]). *If z is a vector indexed by the vertices of G, such that $\sum_{u \in V(G)} z_u = 1$ and $z_u \geq 0$ for each $u \in V(G)$, then*

$$\sum_{uv \in E(G)} z_u z_v \leq \frac{1}{2}\left(1 - \frac{1}{\omega}\right), \tag{4.1}$$

where ω is the clique number of G.

Nikiforov [110] used this lemma in the following way. Let λ_1 and x be the spectral radius and the principal eigenvector of G. We have, by the Rayleigh quotient and the Cauchy-Schwarz inequality,

$$\lambda_1^2 = \left(2 \sum_{uv \in E(G)} x_u x_v\right)^2 \leq 4m \sum_{uv \in E(G)} x_u^2 x_v^2.$$

Since $\sum_{u \in V(G)} x_u^2 = 1$, the Motzkin-Straus lemma directly implies

Theorem 4.1 ([110]). $\lambda_1^2 \leq 2m \left(1 - \frac{1}{\omega}\right)$.

The previous inequality was conjectured earlier by Edwards and Elphick [57].

Further, by the Turán's theorem [151], for any $K_{\omega+1}$-free graph G with n vertices and m edges holds

$$m \leq \frac{n^2}{2} \left(1 - \frac{1}{\omega}\right),$$

so that Theorem 4.1 implies earlier result of Wilf [163].

Corollary 4.1 ([163]). $\lambda_1 \leq n \left(1 - \frac{1}{\omega}\right)$.

Actually, Wilf [163] has applied the Motzkin-Straus lemma directly to the principal eigenvector x, normalized by the sum s of its entries to obtain

$$\lambda_1 \leq s^2 \left(1 - \frac{1}{\omega}\right),$$

of which the above inequality is also a corollary by observing that $s^2 \leq n$.

Let k_i denote the number of all cliques of order i of a graph G. Nikiforov [110] also showed the following

Theorem 4.2 ([110]). $\lambda_1^\omega \leq k_2\lambda_1^{\omega-2}+\cdots+(i-1)k_i\lambda_1^{\omega-i}+\cdots+(\omega-1)k_\omega$.

The proof of this theorem relies on appropriate inequality for the number W_l of all walks of length l in G and $\lim_{l\to\infty} \frac{W_{l+s}}{W_l} = \lambda_1^s$, which follows from the spectral decomposition (1.1). This theorem is also important part of the proof of

Theorem 4.3 ([111]). *The maximum spectral radius among graphs with n vertices and the clique number ω is attained by the Turán graph $T_{n,\omega}$.*

Proof. If G has clique number ω, then

$$\lambda_1(G)^\omega \leq \sum_{i=2}^\omega (i-1)k_i(G)\lambda_1(G)^{\omega-i}$$

by Theorem 4.2. The result of Zykov [169] states that

$$k_s(G) \leq k_s(T_{n,\omega}), \quad 2 \leq s \leq \omega,$$

with equality if and only if $G \cong T_{n,\omega}$. Hence

$$\lambda_1(G)^{\omega} \leq \sum_{i=2}^{\omega} (i-1)k_i(T_{n,\omega})\lambda_1(G)^{\omega-i}.$$

This implies that $\lambda_1(G) \leq z_1$, where z_1 is the largest root of the polynomial

$$z^{\omega} - \sum_{i=2}^{\omega} (i-1)k_i(T_{n,\omega})z^{\omega-i}.$$

As $k_i(T_{n,\omega})$ is the ith elementary symmetric function of the numbers of vertices in distinct parts of $T_{n,\omega}$, we see that the above polynomial is a factor of the characteristic polynomial (3.25) of $T_{n,\omega}$, implying that $z_1 = \lambda_1(T_{n,\omega})$. $\square$

On the other side, Stevanović and Hansen [146] proved the following

Theorem 4.4. *The minimum spectral radius among connected graphs with n vertices and the clique number ω is attained by the kite $KP_{\omega,n-\omega+1}$.*

Proof. Let $\mathcal{G}_{n,\omega}$ be the set of connected graphs with n vertices and the clique number ω.

If $n = \omega$, then $\mathcal{G}_{n,\omega}$ consists of a single graph K_n, which is also the kite $KP_{n,1}$.

If $\omega = 2$, then the path $P_n \cong KP_{2,n-1}$ has the minimum spectral radius among connected graphs with n vertices [38, 98].

Thus, suppose that $n > \omega \geq 3$. We will transform arbitrary $G \in \mathcal{G}_{n,\omega}$, containing a clique K of size ω, into a kite $KP_{\omega,n-\omega+1}$ in a series of steps such that the spectral radius of transformed graph decreases at each step.

Firstly, deletion of an edge from a connected graph strictly decreases its spectral radius by (1.4). In order to keep edge-deleted graph within $\mathcal{G}_{n,\omega}$, we may delete from G any edge not in K which belongs to a cycle. Let G_1 be the subgraph of G obtained by deleting such edges in an arbitrary order as long as they exist. G_1 will, thus, consist of a clique K with a number of rooted trees attached to vertices of K.

Secondly, we will "flatten out" the trees attached to clique vertices. Let T be a rooted tree of G_1, attached to a clique vertex. Let u be the leaf of T farthest from the clique K, and let v be the vertex of T of degree at least 3, closest to u. If T is not a path, then there is a neighbor w of v in T, which is not on the path from v to u. The tree $T - vw + uw$ then has larger distance from the farthest leaf to the clique K than T has, while $\lambda_1(T - vw + uw) < \lambda_1(T)$ by the following

Lemma 4.2. *Let vw be a bridge of a connected graph G and suppose that there exists a path of length k, k ≥ 1, attached at v, with u being the other endpoint of this path. Then*

$$\lambda_1(G - vw + uw) \leq \lambda_1(G)$$

with equality if and only if v has degree 2. □

Processing further with transformations of this type by always choosing farthest leaves in rooted trees, we decrease the spectral radius at each step until we reach a graph G_2 in which every rooted tree, attached to a vertex of K, becomes a path.

The last part of the proof relies on classical lemma of Li and Feng [89] (see also [46, Theorem 6.2.2]).

Lemma 4.3. *Let u and v be two adjacent vertices of a connected graph G and for positive integers k and l, let $G_{k,l}$ denote the graph obtained from G by adding pendant paths of length k at u and length l at v. If k ≥ l ≥ 1, then*

$$\lambda_1(G_{k,l}) > \lambda_1(G_{k+1,l-1}).$$

□

Hence suppose that the graph G_2 consists of the clique K and the paths $P_{k_1}, P_{k_2}, \ldots, P_{k_l}$ attached to l distinct vertices of K, such that $k_1 \geq \cdots \geq k_l$. With repeated use of Lemma 4.3, we may decrease the spectral radius of G_2 until the attached paths $P_{k_2}, \ldots, P_{k_l}$ disappear, and we finally arrive at the kite $KP_{\omega, n-\omega+1}$.

Since we have strictly decreased the spectral radius at each of the previous steps, we conclude that the kite $KP_{\omega, n-\omega+1}$ has smaller spectral radius than any other graph in $\mathcal{G}_{n,\omega}$.

The spectral radius of kites has been estimated by Stevanović and Hansen [146], but better estimates were provided by Cioabă and Gregory in [33]: for $\omega, r \geq 3$ holds

$$\omega - 1 + \frac{1}{\omega(\omega - 1)} < \lambda_1(KP_{\omega,r}) < \omega - 1 + \frac{1}{(\omega - 1)^2}.$$

The best possible upper bound is, of course, given by the spectral radius of the infinite kite IK_ω (see Example 2.3)

$$\lambda_1(KP_{\omega,r}) < \lambda_1(IK_\omega) = \frac{\omega - 3}{2} + \frac{\omega - 1}{2(\omega - 2)}\sqrt{\omega^2 - 4}$$

$$\approx \omega - 1 + \frac{1}{\omega(\omega - 2)} - \frac{1}{\omega^3}\frac{\omega - 1}{\omega - 2}.$$

4.3 CHROMATIC NUMBER

Relations between the spectral radius λ_1 and the chromatic number χ have a long history. Wilf [162] proved in 1967 that

$$\chi \leq \lambda_1 + 1. \tag{4.2}$$

Recall that a graph G is color critical if the removal of any vertex of G lowers its chromatic number. Every graph contains a color-critical induced subgraph, and Wilf used the fact that in a χ-colorable color-critical graph the degree of each vertex is at least $\chi - 1$, so that by the monotonicity of spectral radius, $\lambda_1 \geq \chi - 1$.

Cvetković [42] in 1972 and, independently, Edwards and Elphick [57] in 1983 proved that

$$\chi \geq \frac{n}{n - \lambda_1}$$

or, equivalently,

$$\lambda_1 \leq n\left(1 - \frac{1}{\chi}\right). \tag{4.3}$$

These authors observed that, for fixed chromatic number χ, the edge-maximal χ-colorable graph is the complete multipartite graph $K_{n_1,\ldots,n_\chi}$ for some parameters $n_1, \ldots, n_\chi$ with $n_1 + \cdots + n_\chi = n$. To finish the proof of (4.3), Cvetković used the formula (3.25) for the characteristic polynomial of complete multipartite graphs and properties of elementary symmetric functions, while Edwards and Elphick [57] relied on the formula (3.22):

$$x_i = \frac{\sum_{k=1}^{\chi} n_k x_k}{\lambda_1 + n_i}$$

for the principal eigenvector components in part n_i of $K_{n_1,\ldots,n_\chi}$. Assume, without loss of generality, that $n_1 \leq \cdots \leq n_\chi$ so that $x_1 \geq \cdots \geq x_\chi$. Then by the Chebyshev's sum inequality

$$\sum_{k=1}^{\chi} n_k x_k \leq \frac{\sum_{k=1}^{\chi} n_k \sum_{k=1}^{\chi} x_k}{\chi} = \frac{n}{\chi} \sum_{k=1}^{\chi} x_k.$$

Further, (3.22) yields

$$\lambda_1 \sum_{i=1}^{\chi} x_i = (\chi - 1) \sum_{k=1}^{\chi} n_k x_k \leq n \left(1 - \frac{1}{\chi} \right) \sum_{k=1}^{\chi} x_k,$$

which implies (4.3). Certainly, (4.3) is also implied by Corollary 4.1 by noticing that $\omega \leq \chi$.

As for the graphs with extremal spectral radii among χ-colorable graphs, the maximum spectral radius is attained by the Turán graph $T_{n,\chi}$ by Theorem 3.25. Graphs with the minimum spectral radius have been characterized by Feng et al. [59].

Theorem 4.5 ([59]). *The χ-colorable connected graph with the minimum spectral radius is*

1) *the path P_n, if $\chi = 2$;*
2) *the cycle C_n, if $\chi = 3$ and n is odd;*
3) *the lollipop $CP_{n-1,2}$ to which a pendant vertex is added, if $\chi = 3$ and n is even;*
4) *the kite $KP_{\chi,n-\chi+1}$, if $\chi \geq 4$.*

Proof. Part (1) is straightforward, as the path P_n has the minimum spectral radius not only among bipartite graphs, but also among all connected graphs [38, 98].

Parts (2) and (3) rely on the internal path Lemma 1.4. If G is a 3-colorable graph, then it contains an odd cycle C_h as a color-critical subgraph. If $h = n$, so that n is odd as well, then (b) holds. If $h < n$, then G contains the lollipop $CP_{h,2}$, which is the cycle C_h with a pendant edge attached, as a subgraph.

If n is even, then by inserting $n - h - 1$ vertices on the unique internal path of $CP_{h,2}$, we get from Lemma 1.4 that

$$\lambda_1(G) \geq \lambda_1(CP_{h,2}) > \lambda_1(CP_{n-1,2}),$$

proving (3). If n is odd, then by inserting $n - h$ vertices on the internal path of $CP_{h,2}$, we get that

$$\lambda_1(G) \geq \lambda_1(CP_{h,2}) > \lambda_1(CP_{n,2}) > \lambda_1(C_n),$$

proving (2).

For part (4), if G contains the complete graph K_χ as a color-critical subgraph, then by Theorem 4.4

$$\lambda_1(G) \geq \lambda_1(KP_{\chi,n-\chi+1}).$$

Hence suppose that for the color-critical subgraph H of G holds $H \not\cong K_\chi$. The following result of Krivelevich [86] gives a lower bound on the number of edges of H.

Lemma 4.4. *If H is a χ-colorable color-critical graph, such that $H \not\cong K_\chi$, then*

$$|E(H)| \geq \left(\frac{\chi - 1}{2} + \frac{\chi - 3}{2(\chi - 1)^2} \right) |V(H)|.$$

$\square$

Thus, we have

$$\lambda_1(G) \geq \lambda_1(H) \geq \frac{2|E(H)|}{|V(H)|} = \chi - 1 + \frac{\chi - 3}{(\chi - 1)^2}.$$

It is now a matter of elementary transformations to show that

$$\chi - 1 + \frac{\chi - 3}{(\chi - 1)^2} > \frac{\chi - 3}{2} + \frac{\chi - 1}{2(\chi - 2)} \sqrt{\chi^2 - 4}$$
$$= \lambda_1(IK_\chi) > \lambda_1(KP_{\chi,n-\chi+1}),$$

where IK_χ is the infinite kite (see Example 2.3).

4.4 INDEPENDENCE NUMBER

Due to the well-known inequality $\alpha \geq \frac{n}{\chi}$ between the independence and the chromatic number, (4.2) immediately yields

$$\alpha \ge \frac{n}{\lambda_1 + 1}.$$

Wilf [163] has further improved this lower bound in the case of regular graphs by introducing into consideration the smallest eigenvalue λ_n and its eigenvector x_n. Specifically, let

$$M_+ = \min_{(x_n)_i > 0} \frac{1}{(x_n)_i}, \qquad M_- = \min_{(x_n)_i < 0} \frac{1}{|(x_n)_i|}.$$

Wilf has shown that

Theorem 4.6 ([163]). *If G is a r-regular graph with n vertices, then*

$$\alpha \ge \frac{n}{r + 1 + \frac{\lambda_n+1}{n} \max\{M_+^2, M_-^2\}}. \tag{4.4}$$

Wilf has argued that (4.4) is a good bound if G admits as an eigenvector a vector whose all entries are equal in absolute value, and posed the following interesting

Question 4.1([163]). What kind of a graph can have an eigenvector consisting solely of ± 1 entries?

It is straightforward to determine the graph G with the maximum spectral radius among graphs with independence number α. Simply, let S be an independent set of size α in G. Due to the monotonicity of spectral radius (1.4), G has the maximum spectral radius if and only if it contains all edges between vertices in S and vertices in $V(G) \setminus S$, and also edges between all pairs of vertices in $V(G) \setminus S$, or in other words, if and only if G is a complete split graph $CS_{n,n-\alpha}$.

With respect to the minimum spectral radius, if we allow disconnected graphs, then the minimum spectral radius among graphs with n vertices and independence number α is obtained for a union of α cliques, the size of each being either $\lfloor n/\alpha \rfloor$ or $\lceil n/\alpha \rceil$. However, determining the graph with the minimum spectral radius among connected graphs with independence number α appears to be a tough problem. So far, such graphs are known only for a few values of α: Xu et al. [165] determined them for $\alpha \in \{1, 2, \lceil n/2 \rceil, \lceil n/2 \rceil + 1, n-3, n-2, n-1\}$, while Du and Shi [56] determined them for $\alpha \in \{3, 4\}$, provided that $\alpha | n$.

The only graph with $\alpha = 1$ is the complete graph K_n, while the only connected graph with $\alpha = n - 1$ is the star $K_{1,n-1}$, so that the problem is

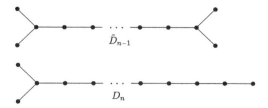

Figure 4.1 The graphs $\tilde{D}_{n-1}$ and D_n.

trivial for these values. Further, the path P_n, as the connected graph with the minimum spectral radius, has $\alpha(P_n) = \lceil n/2 \rceil$, settling this case as well. The case $\alpha = \lceil n/2 \rceil + 1$ has been settled by considering the Smith graphs [137]— the set of graphs whose spectral radius is at most 2. The minimum spectral radius in such case is attained by $\tilde{D}_{n-1}$ if n is odd, and by D_n if n is even (see Fig. 4.1; both graphs have n vertices, regardless of the subscript).

Let $H_3(n_1, n_2)$ be the graph obtained from the path P_3 by attaching n_1 pendant vertices to one endpoint, and n_2 pendant vertices to the other endpoint of P_3. Similary, let $H_5(n_1, n_2, n_3)$ be the graph obtained from the path P_5 by attaching n_1 pendant vertices to one endpoint, n_2 pendant vertices to the middle vertex, and n_3 pendant vertices to the other endpoint of P_5. Xu et al. [165] have shown that $H_3(\lfloor (n-3)/2 \rfloor, \lceil (n-3)/2 \rceil)$ has the minimum spectral radius among connected graphs with $\alpha = n - 2$, while $H_5(\lfloor (n-2)/3 \rfloor, n-2\lfloor (n-2)/3 \rfloor - 5, \lfloor (n-2)/3 \rfloor)$ has the minimum spectral radius among connected graphs with $\alpha = n - 3$.

Further, for natural numbers $n_1, \ldots, n_k \geq 2$, let $F(n_1, \ldots, n_k)$ denote a chain of cliques, obtained from the union $K_{n_1} \cup \cdots \cup K_{n_k}$ by joining with an edge a vertex of K_{n_1} to a vertex of K_{n_2}, a vertex of K_{n_2} to a vertex of K_{n_3}, and so on until we reach K_{n_k} (without joining the same vertex to two vertices in different cliques). Relying on Turán's bound on the number of edges in K_{r+1}-free graphs [151], Xu et al. [165] have shown that $F(\lfloor n/2 \rfloor, \lceil n/2 \rceil)$ has the minimum spectral radius for $\alpha = 2$. Du and Shi [56] have further shown that $F(n/3, n/3, n/3)$ has the minimum spectral radius for $\alpha = 3$ and $3|n$, while $F(n/4, n/4, n/4, n/4)$ has the minimum spectral radius for $\alpha = 4$ and $4|n$. In addition, Du and Shi have conjectured that

Conjecture 4.4 ([56]). *For each $\alpha \in \mathbb{N}$ there exists $n_0 \in \mathbb{N}$ such that for each $n > n_0$, the minimum spectral radius among connected graphs with n vertices and the independence number α is attained by a graph $F(n_1, \ldots, n_\alpha)$, for which $n_i \in \{\lfloor n/\alpha \rfloor, \lceil n/\alpha \rceil\}$ for each $i = 1, \ldots, \alpha$.*

This conjecture looks very plausible. If G is an arbitrary connected graph with n vertices and the independence number α, then its complement $\overline{G}$ has the clique number α. By the Turan's theorem [151], $\overline{G}$ then has at most as many edges as $T_{n,\alpha}$, that is

$$\binom{n}{2} - \sum_{i=1}^{\alpha} \binom{n_i}{2}$$

for $n_i \in \{\lfloor n/\alpha \rfloor, \lceil n/\alpha \rceil\}$ and $\sum_{i=1}^{\alpha} n_i = n$. Hence, G itself has at least $\sum_{i=1}^{\alpha} \binom{n_i}{2}$ edges and its average degree, discarding $\lfloor \rfloor$ and $\lceil \rceil$ for the moment, is at least

$$\frac{2\sum_{i=1}^{\alpha} \binom{n_i}{2}}{n} = \frac{\sum_{i=1}^{\alpha} n_i^2}{n} - 1 \approx \frac{n}{\alpha} - 1.$$

Thus,

$$\frac{n}{\alpha} - 1 \le \lambda_1(G).$$

On the other hand, the maximum vertex degree of $F(n_1, \ldots, n_\alpha)$ is $\lceil n/\alpha \rceil$ (assuming that α does not divide n), so that

$$\lambda_1(F(n_1, \ldots, n_\alpha)) \le \lceil n/\alpha \rceil.$$

Therefore, the spectral radius of $F(n_1, \ldots, n_\alpha)$ is very close to the lower bound, making it a likely candidate for the minimum spectral radius. It is to be expected that candidate graphs for the minimum spectral radius, in general, consist of α cliques, each of order $\lfloor n/\alpha \rfloor$ or $\lceil n/\alpha \rceil$, joined together in a tree-like structure in order to ensure connectedness. The path-like ordering of cliques in $F(n_1, \ldots, n_\alpha)$ then once more suggests that it should have the minimum spectral radius among such candidate graphs.

4.5 MATCHING NUMBER

Graphs with the maximum spectral radius among graphs with n vertices and the matching number ν have been characterized by Feng et al. [61].

Theorem 4.7 ([61]). *The maximum spectral radius among graphs with n vertices and the matching number ν is attained by*

1) K_n if $n = 2\nu$ or $n = 2\nu + 1$;
2) $K_{2\nu+1} \cup \overline{K_{n-2\nu-1}}$ if $2\nu + 2 \le n \le 3\nu + 1$;

3) $K_{2v+1} \cup \overline{K_{n-2v-1}}$ and $K_v \vee \overline{K_{n-v}}$ if $n = 3v + 2$;
4) $K_v \vee \overline{K_{n-v}}$ if $3v + 3 \leq n$.

The maximum graphs in the previous theorem are disconnected in case (2). It does not seem to be known what connected graphs have the maximum spectral radius if $2v + 2 \leq n \leq 3v + 1$.

The proof of the above theorem starts with the Tutte-Berge formula [13, 152] stating that the size of a maximum matching of a graph $G = (V, E)$ equals

$$v = \frac{1}{2} \min_{S \subseteq V} (|V| + |S| + \text{odd}(G - U)),$$

where $\text{odd}(G - U)$ is the number of components in $G - U$ with an odd number of vertices. Let S be the subset on s vertices such that $G - S$ has $q = n + s - 2v$ odd components $G_1, \ldots, G_q$. Let $n_i = |V(G_i)|$, $i = 1, \ldots, q$ and assume, without loss of generality, that $n_1 \leq \cdots \leq n_q$. Since S and $G_1, \ldots, G_q$ are disjoint, clearly $n \geq s + q = n + 2s - 2v$, so that $s \leq v$.

Since the spectral radius is monotone with respect to addition of edges, we see that in the graph G with the maximum spectral radius each of S and $G_1, \ldots, G_q$ has to induce a complete graph, and that it also has to contain all edges between a vertex in S and a vertex in $\cup_{i=1}^{q} G_i$, showing that

$$G \cong K_s \vee \cup_{i=1}^{q} K_{n_i}.$$

Next, let λ_1 and x denote the spectral radius and the principal eigenvector of G. All components of x within K_{n_i} are equal to each other, so that we will denote them shortly as y_i. Similarly, the common component of x in K_s is denoted as y_0. From the eigenvalue equation we have

$$\lambda_1 y_0 = (s - 1)y_0 + \sum_{i=1}^{q} n_i y_i,$$
$$\lambda_1 y_1 = (n_1 - 1)y_1 + s y_0,$$
$$\cdots \quad \cdots$$
$$\lambda_1 y_q = (n_q - 1)y_q + s y_0.$$

Expressing $y_1, \ldots, y_q$ in terms of y_0 and replacing them in the first equation, we see that λ_1 satisfies

$$\frac{\lambda_1 - s + 1}{s} - \sum_{i=1}^{q} \frac{n_i}{\lambda_1 - n_i + 1} = 0. \tag{4.5}$$

Let us define the following modification of the left-hand side above as

$$f(\lambda, k) = \frac{\lambda - s + 1}{s} - \sum_{i=1}^{q-2} \frac{n_i}{\lambda - n_i + 1} + \frac{n_{q-1} - k}{\lambda - (n_{q-1} - k) + 1}$$

$$+ \frac{n_q + k}{\lambda - (n_q + k) + 1}.$$

Hence, the spectral radius of $K_s \vee \cup_{i=1}^{q} K_{n_i}$ is the largest root of the equation $f(\lambda, 0) = 0$, while the spectral radius of $K_s \vee \left(\cup_{i=1}^{q-2} K_{n_i} \cup K_{n_{q-1}-k} \cup K_{n_q+k} \right)$ is the largest root of the equation $f(\lambda, k) = 0$.

Recalling that $n_1 \le \cdots \le n_{q-1} \le n_q$, we now show that if $3 \le n_{q-1}$, then for $G \cong K_s \vee \cup_{i=1}^{q} K_{n_i}$ and $G' = K_s \vee \left(\cup_{i=1}^{q-2} K_{n_i} \cup K_{n_{q-1}-2} \cup K_{n_q+2} \right)$ holds $\lambda_1(G) < \lambda_1(G')$. Derivating with respect to k, we obtain for $\lambda \ge n_q - 1$,

$$\frac{\partial f}{\partial k} = \frac{\lambda + 1}{(\lambda - n_{q-1} + 1 + k)^2} - \frac{\lambda + 1}{(\lambda - n_q + 1 - k)^2}$$

$$= \frac{(\lambda + 1)(n_{q-1} - n_q - 2k)(2\lambda - n_{q-1} - n_q - 2)}{(\lambda - n_{q-1} + 1 + k)^2(\lambda - n_q + 1 - k)^2}$$

$$< 0,$$

showing that $f(\lambda, k)$ is strictly decreasing with respect to k for $\lambda \ge n_q - 1$. Since G contains K_{n_q}, we have $\lambda_1(G) \ge n_q - 1$ (and also $\lambda_1(G') \ge n_q - 1$ as G' also contains K_{n_q}), so that

$$0 = f(\lambda_1(G), 0) > f(\lambda_1(G), 2).$$

Further, derivating $f(\lambda, 2)$ with respect to λ, we see that for $\lambda \ge n_q - 1$

$$\frac{\partial f(\lambda, 2)}{\partial \lambda} = \frac{1}{s} + \sum_{i=1}^{q-2} \frac{n_i}{(\lambda - n_i + 1)^2} + \frac{n_{q-1} - 2}{(\lambda - n_{q-1} + 3)^2} + \frac{n_q + 2}{(\lambda - n_q - 1)^2} > 0,$$

showing that $f(\lambda, 2)$ is strictly increasing for $\lambda \ge n_q - 1$. Since $\lambda_1(G')$ is the largest root of the equation $f(\lambda, 2) = 0$, while $f(\lambda_1(G), 2) < 0$, we conclude that $\lambda_1(G) < \lambda_1(G')$, a contradiction as G is assumed to have the maximum spectral radius.

Hence $n_1 = \cdots = n_{q-1} = 1$ and $n_q = 2\nu - 2s + 1$, so that the maximum graph has to be sought in the form

$$G \cong K_s \vee \left(\cup_{i=1}^{q-1} K_1 \cup K_{2\nu-2s+1} \right) \cong K_s \vee \left(\overline{K_{q-1}} \cup K_{2\nu-2s+1} \right).$$

The rest of the proof in [61] consists of a somewhat tedious estimation of the largest root of (4.5), depending on the cases stated in Theorem 4.7.

Characterization of graphs with the minimum spectral radius among connected graphs with the matching number ν is an open problem. The only result in this direction so far is the following lower bound of Stevanović [139].

Theorem 4.8 ([139]). *If G has m edges and the matching number ν, then*

$$\lambda_1 \geq \sqrt{\left\lceil \frac{m+\nu}{2\nu} \right\rceil}.$$

Proof. Let M be a matching of G with ν edges. Since M cannot be extended to a larger matching, each edge $e \notin M$ is incident to one or two edges in M. Thus, the sum of degrees of 2ν endpoints of edges in M is at least $m + \nu$, where 2μ comes from the edges in M and $m - \nu$ is the contribution of edges not in M. Therefore, at least one of these 2ν vertices has degree at least $\lceil \frac{m+\nu}{2\nu} \rceil$, so that for the maximum vertex degree Δ holds

$$\Delta \geq \left\lceil \frac{m+\nu}{2\nu} \right\rceil.$$

The statement now follows from the well-known result $\lambda_1 \geq \sqrt{\Delta}$, as $\sqrt{\Delta}$ is the spectral radius of the star $K_{1,\Delta}$ contained in G. $\square$

4.6 THE DIAMETER

We will cover here a number of results on the spectral radius of graphs with a given diameter. Firstly, we deal with the graphs having the maximum spectral radius. Infinite bugs have already been considered in Example 2.4. We first recall the definition of its finite counterpart.

Definition 4.1. A *bug* Bug_{p,q_1,q_2} is a graph obtained from a complete graph K_p by deleting an edge uv and identifying the endpoints of paths P_{q_1}

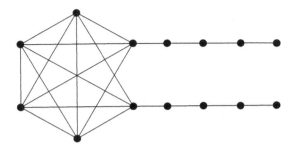

Figure 4.2 A $Bug_{6,5,5}$.

and P_{q_2} with u and v, respectively. A bug is *balanced* if $|q_1 - q_2| \leq 1$ (see Fig. 4.2).

The number of vertices in Bug_{p,q_1,q_2} is $n = p + q_1 + q_2 - 2$ and the number of edges is $m = \binom{p}{2} + q_1 + q_2 - 3$.

Experiments with AutoGraphiX [27] led to a conjecture that a balanced bug has the maximum spectral radius among graphs with a given diameter, which was then proved by Hansen and Stevanović in [72]. The same result was proved independently by van Dam [153]. It is an interesting twist of fate that [72] was submitted to a special issue of a journal about a year before and [153] was submitted to a standard issue of a journal, however, as the preparations of a special issue took the full three years, [72] was published about a year later than [153] as the end result.

Theorem 4.9 ([72, 153]). *Among all graphs with n vertices and diameter D, the maximum spectral radius is attained by*

1) a complete graph K_n when $D = 1$, and
2) a balanced bug $Bug_{n-D+2,\lceil D/2 \rceil,\lfloor D/2 \rfloor}$ when $D \geq 2$.

We will closely follow the proof from [72] here.

Proof. A complete graph K_n is the unique graph with n vertices and diameter 1, so the theorem holds for $D = 1$. A path P_n is the unique graph with n vertices and diameter $n - 1$, and since P_n is isomorphic to $Bug_{3,i,n-i-1}$ for $1 \leq i \leq n - 2$, the theorem holds for $D = n - 1$.

Now suppose that $2 \leq D \leq n - 2$. Let $G = (V, E)$ be a graph with maximum spectral radius among all graphs with n vertices and diameter D.

Let c be a vertex of G with eccentricity D. Denote by L_i the set of vertices at distance i from c and let $l_i = |L_i|$, $i = 0, 1, \ldots, D$.

Since the spectral radius of a graph increases by adding an edge, it follows that G contains all edges between vertices of L_i, $i = 1, 2, \ldots, D$, and all edges between vertices of L_{i-1} and L_i, $i = 1, 2, \ldots, D$.

A bug $Bug_{n-D+2,i,D-i}$ with diameter D, $i = 1, 2, \ldots, D - 1$, contains the graph $K_{n-D+2} - e$ as its induced subgraph. It is straightforward to see that

$$\lambda_1 (K_{n-D+2} - e) = \frac{1}{2} \left(n - D - 1 + \sqrt{(n - D + 2)^2 + 2(n - D + 2) - 7} \right),$$

which is strictly larger than $n - D$. By the Interlacing theorem it follows that

$$\lambda_1 (Bug_{n-D+2,i,D-i-1}) > n - D.$$

Thus, G as a graph with maximum spectral radius must satisfy $\lambda_1(G) > n - D$. The maximum vertex degree Δ of G is larger than or equal to $\lambda_1(G)$, so that

$$\Delta \geq n - D + 1.$$

On the other hand, it holds that

$$\Delta = \max\{l_1\} \cup \{l_{i-1} + l_i + l_{i+1} - 1 : i = 1, 2, \ldots, D - 1\} \cup \{l_{D-1} + l_D - 1\}.$$

Since $l_i \geq 1$ and $\sum_{i=1}^{D} l_i = n - 1$, it follows that $\Delta \geq n - D + 1$ is possible if and only if for some $1 \leq j \leq D - 1$ it holds that

$$l_{j-1} + l_j + l_{j+1} = n - D + 2, \quad l_1 = \cdots = l_{j-2} = 1, \quad \text{and}$$
$$l_{j+2} = \cdots = l_D = 1.$$

So, we have that $\Delta = n - D + 1$ and $n - D < \lambda_1(G) \leq n - D + 1$.

In order to show that G is a bug, we still have to show that two of l_{j-1}, l_j and l_{j+1} are equal to 1. We will heavily resort to the increase of the spectral radius by edge rotations for this (see Lemma 1.1). Since the components of principal eigenvector x corresponding to similar vertices are equal, we may denote by x_i the component of x corresponding to each vertex in L_i, $i = 0, 1, \ldots, D$.

Let s be the unique vertex of L_{j-2} and let t be a vertex in L_{j+1}. If $l_{j-1} > 1$, moving an arbitrary vertex r from L_{j-1} to L_j results in the rotation of the edge rs to the edge rt, followed by addition of edges between r and the remaining vertices in L_{j+1}. Therefore, if $x_{j-2} \le x_{j+1}$ we may move a vertex from L_{j-1} to L_j and increase the spectral radius, while keeping the diameter intact. Since G has the maximum spectral radius, this is a contradiction. Thus, either $l_{j-1} = 1$ or $x_{j-2} > x_{j+1}$. Similarly, by considering the unique vertex of L_{j+2} we conclude that either $l_{j+1} = 1$ or $x_{j+2} > x_{j-1}$.

Now we show that it is impossible that $x_{j-1} > x_j < x_{j+1}$. In such a case, the eigenvalue equation gives

$$(\lambda_1(G)+1)x_j = l_j x_j + l_{j-1} x_{j-1} + l_{j+1} x_{j+1} > (l_{j-1} + l_j + l_{j+1})x_j = (n-D+2)x_j,$$

from where it follows that $\lambda_1(G) > n - D + 1$, which is a contradiction. Therefore, we have that either $x_{j-1} \le x_j$ or $x_j \ge x_{j+1}$.

Without loss of generality, suppose that $x_{j-1} \le x_j$. If $l_{j-1} > 1$, let G' be a graph obtained by moving an arbitrary vertex of L_{j-1} to L_j. Using the Rayleigh quotient, we have that

$$\lambda_1(G') = \max_{y \ne 0} \frac{y^T A(G')y}{y^T y} \ge \frac{x^T A(G')x}{x^T x}.$$

Further, we have that

$$x^T A(G')x = x^T A(G)x + 2(l_{j+1}x_{j-1}x_{j+1} - x_{j-2}x_{j-1}).$$

From the eigenvalue equation, we have that

$$(\lambda_1(G) + 1)x_{j-1} = x_{j-2} + l_{j-1}x_{j-1} + l_j x_j,$$
$$(\lambda_1(G) + 1)x_j = l_{j+1}x_{j+1} + l_{j-1}x_{j-1} + l_j x_j,$$

from where it follows that

$$l_{j+1}x_{j+1} - x_{j-2} = (\lambda_1(G) + 1)(x_j - x_{j-1}).$$

Then

$$x^T A(G')x = x^T A(G)x + 2x_{j-1}(\lambda_1(G) + 1)(x_j - x_{j-1}) \ge x^T A(G)x,$$

and so

$$\lambda_1(G') \ge \frac{x^T A(G)x}{x^T x} = \lambda_1(G).$$

However, equality cannot hold here. For suppose that $\lambda_1(G') = \lambda_1(G)$: then we must have that $x_j = x_{j-1}$ and that x is an eigenvector of G' corresponding to $\lambda_1(G') = \lambda_1(G)$. The eigenvalue equation for a vertex of L_{j-2} gives in G

$$\lambda_1(G)x_{j-2} = x_{j-3} + l_{j-1}x_{j-1},$$

while in G' it gives

$$\lambda_1(G)x_{j-2} = x_{j-3} + (l_{j-1} - 1)x_{j-1},$$

implying that $x_{j-1} = 0$, which is impossible, as x is a positive eigenvector. Thus, it must hold that $\lambda_1(G') > \lambda_1(G)$, which is a contradiction with the choice of G. Therefore, it must hold that $l_{j-1} = 1$.

We are now half the way done. We have shown that one of l_{j-1}, l_j, and l_{j+1} (actually, l_{j-1}) is equal to 1. If one of l_j and l_{j+1} is equal to 1, we have a bug.

Otherwise, suppose that $l_j > 1$ and $l_{j+1} > 1$. Moving $l_j - 1$ vertices from L_j to L_{j+1} in G results in a graph G'', whose adjacency matrix $A(G'')$ satisfies

$$x^T A(G'')x = x^T A(G)x + 2(l_j - 1)x_j(x_{j+2} - x_{j-1}).$$

Moving $l_{j+1} - 1$ vertices from L_{j+1} to L_j in G results in a graph G''', whose adjacency matrix $A(G''')$ satisfies

$$x^T A(G''')x = x^T A(G)x - 2(l_{j+1} - 1)x_{j+1}(x_{j+2} - x_{j-1}).$$

Thus, if $x_{j+2} \neq x_{j-1}$, one of the graphs G'' and G''' has spectral radius larger than $\lambda_1(G)$, which is a contradiction. If $x_{j+2} = x_{j-1}$, then from the eigenvalue equation

$$(\lambda_1(G) + 1)x_j = l_jx_j + l_{j+1}x_{j+1} + x_{j-1},$$
$$(\lambda_1(G) + 1)x_{j+1} = l_jx_j + l_{j+1}x_{j+1} + x_{j+2},$$

we get that $x_j = x_{j+1}$. However, in order that, say, G'' has spectral radius equal to $\lambda_1(G)$, x must be an eigenvector of G''. The eigenvalue equations for a vertex of L_{j-1} in G and G'', respectively, are

$$\lambda_1(G)x_{j-1} = x_{j-2} + l_jx_j,$$
$$\lambda_1(G)x_{j-1} = x_{j-2} + x_j,$$

showing that $x_j = 0$ (as we supposed that $l_j > 1$), which is a contradiction. Thus, one of l_j and l_{j+1} must also be equal to 1, and we conclude that G is indeed a bug of the form Bug_{n-D+2,q_1,q_2}. Let $p = n - D + 2$ in the rest of the proof.

It remains to show that the bug G is balanced, i.e., that $|q_1 - q_2| \leq 1$. We show that, whenever $q_1 \geq q_2 \geq 1$, it holds that

$$\lambda_1(Bug_{p,q_1,q_2}) > \lambda_1(Bug_{p,q_1+1,q_2-1}), \tag{4.6}$$

implying that the largest spectral radius is obtained when $|q_1 - q_2| \leq 1$. As the largest eigenvalue of bugs cannot be explicitly calculated (unless we have an infinite bug—see Example 2.4), we now turn to the characteristic polynomials of bugs. In order to prove (4.6) it is enough to show that

$$(\forall x \geq \lambda_1(Bug_{p,q_1,q_2})) \quad P_{Bug_{p,q_1,q_2}}(x) < P_{Bug_{p,q_1+1,q_2-1}}(x), \tag{4.7}$$

as the previous inequality implies that $P_{Bug_{p,q_1+1,q_2-1}}(x)$ cannot have real roots that are greater than or equal to $\lambda_1(Bug_{p,q_1,q_2})$.

Consider the polynomial

$$W_{q_1,q_2}(x) = P_{Bug_{p,q_1,q_2}}(x) - P_{Bug_{p,q_1+1,q_2-1}}(x).$$

It was proved in [131] that, if an edge $e = \{u, v\}$ is a pending edge of G, with v being a vertex of degree 1, then

$$P_G(x) = xP_{G-v}(x) - P_{G-u-v}(x). \tag{4.8}$$

Applying (4.8) first to the pending edge of P_{q_2} in Bug_{p,q_1,q_2} and then to the pending edge of P_{q_1+1} in Bug_{p,q_1+1,q_2-1}, we get that

$$P_{Bug_{p,q_1,q_2}}(x) = xP_{Bug_{p,q_1,q_2-1}}(x) - P_{Bug_{p,q_1,q_2-2}}(x),$$
$$P_{Bug_{p,q_1+1,q_2-1}}(x) = xP_{Bug_{p,q_1,q_2-1}}(x) - P_{Bug_{p,q_1-1,q_2-1}}(x).$$

Subtracting the above two equalities, it follows that

$$W_{q_1,q_2}(x) = W_{q_1-1,q_2-1}(x).$$

Iterative application of the last equality produces a chain of equalities with smaller values of q_1 and q_2, implying that

$$W_{q_1,q_2}(x) = W_{q_1-q_2+1,1}(x).$$

It was also proved in [131] that, if a vertex v disconnects G into two subgraphs G_1 and G_2, such that v belongs to both of them, then

$$P_G(x) = P_{G_1-v}(x)P_{G_2}(x) + P_{G_1}(x)P_{G_2-v}(x) - xP_{G_1-v}(x)P_{G_2-v}(x). \tag{4.9}$$

Applying (4.9) first to a vertex of $Bug_{p,q_1-q_2+1,1}$ at which $K_p - e$ and $P_{q_1-q_2+1}$ meet, and then to a vertex of $Bug_{p,q_1-q_2+2,0}$ at which K_{p-1} and $P_{q_1-q_2+2}$ meet, we get that

$$P_{Bug_{p,q_1-q_2+1,1}}(x) = P_{K_{p-1}}(x)P_{P_{q_1-q_2+1}}(x) + P_{K_p-e}(x)P_{P_{q_1-q_2}}(x)$$
$$-xP_{K_{p-1}}(x)P_{P_{q_1-q_2}}(x),$$
$$P_{Bug_{p,q_1-q_2+2,0}}(x) = P_{K_{p-1}}(x)P_{P_{q_1-q_2+1}}(x) + P_{K_{p-2}}(x)P_{P_{q_1-q_2+2}}(x)$$
$$-xP_{K_{p-2}}(x)P_{P_{q_1-q_2+1}}(x)$$
$$= P_{K_{p-1}}(x)P_{P_{q_1-q_2+1}}(x) - P_{K_{p-2}}(x)P_{P_{q_1-q_2}}(x),$$

where in the second equality we used that

$$P_{P_{q_1-q_2+2}}(x) = xP_{P_{q_1-q_2+1}}(x) - P_{P_{q_1-q_2}}(x).$$

Thus,

$$W_{q_1-q_2+1,1}(x) = P_{P_{q_1-q_2}}(x)\left(P_{K_p-e}(x) - xP_{K_{p-1}}(x) + P_{K_{p-2}}(x)\right).$$

Now, we show that (4.7) indeed holds. Let $x \geq \lambda_1(Bug_{p,q_1,q_2}) > p \geq 2$. We have that $P_{P_{q_1-q_2}}(x) > 0$, as the largest eigenvalue of a path $P_{q_1-q_2}$ is strictly less than 2. Thus, we have to show that

$$P_{K_p-e}(x) - xP_{K_{p-1}}(x) + P_{K_{p-2}}(x) < 0.$$

From the eigenvalue equations for $K_p - e$ one easily gets that the spectrum of $K_p - e$ consists of simple eigenvalues $\frac{1}{2}\left(p - 3 \pm \sqrt{p^2 + 2p - 7}\right)$, simple eigenvalue 0 and an eigenvalue -1 of multiplicity $p - 3$. Therefore

$$P_{K_p-e}(x) = x(x^2 - (p-3)x - 2(p-2))(x+1)^{p-3}.$$

Thus,

$$P_{K_p-e}(x) - xP_{K_{p-1}}(x) + P_{K_{p-2}}(x)$$
$$= (x+1)^{p-3}\left(x(x^2-(p-3)x-2(p-2)) - x(x-p+2)(x+1)\right.$$
$$\left.+(x-p+3)\right)$$
$$= -(x+1)^{p-3}((p-1)x-(p-3)).$$

Since $p = n - D + 2 \geq 4$, the above expression is less than 0 for all $x > \frac{p-3}{p-1}$, which shows that (4.7) is satisfied, as $\lambda_1(Bug_{p,q_1,q_2}) > p - 2 \geq 2$. $\square$

Even if we restrict ourselves to more specialized graph classes, graphs with the maximum spectral radius for a given diameter continue to be bug-like, with majority of edges concentrated in dense, central core to which two almost equally long paths are attached.

Theorem 4.10 ([135]). *The maximum spectral radius among trees with n vertices and the diameter D is attained by a tree formed from the path P_{D+1} by attaching $n - D - 1$ pendant vertices to a middle vertex of P_{D+1}.*

Note that Simić et al. [135] have, in addition, characterized the graph with the maximum spectral radius in the set of caterpillars with a given degree sequence (which determines both the number of vertices and the diameter).

Theorem 4.11 ([94]). *The maximum spectral radius among unicyclic graphs with n vertices and the diameter D is attained by a graph formed from a triangle by attaching $n - D - 2$ pendant vertices and a path of length $\lfloor D/2 \rfloor$ to one vertex of the triangle, and a path of length $\lceil D/2 \rceil - 1$ to another vertex of the triangle.*

Theorem 4.12 ([167]). *The maximum spectral radis among bipartite graphs with n vertices and the diameter D is attained by a graph formed from the complete bipartite graph $K_{\lceil \frac{n-D+3}{2} \rceil, \lfloor \frac{n-D+3}{2} \rfloor}$ by first removing an edge uv from it, where u belongs to the $\lceil \frac{n-D+3}{2} \rceil$ part and v belongs to the $\lfloor \frac{n-D+3}{2} \rfloor$ part, and then by identifying an endpoint of the path $P_{\lceil \frac{D-1}{2} \rceil}$ with u, and by identifying an endpoint of the path $P_{\lfloor \frac{D-1}{2} \rfloor}$ with v.*

Examples of maximum graphs from Theorems 4.10–4.12 are shown in Fig. 4.3.

Determining graphs with the minimum spectral radius among graphs with a given diameter attracted a lot of research activity recently. This question was motivated in [154] by its application to communication networks, as one wants to minimize both their diameter, in order for messages to travel shorter between vertices, and their spectral radius, in order to make them less prone to virus propagation.

For small diameter, Cioabă et al. [36] have proved the following

Theorem 4.13 ([36]). *If G is a graph with n vertices and diameter D, then*

$$\lambda_1 \geq (n - 1)^{1/D} \tag{4.10}$$

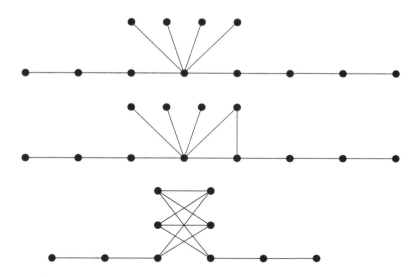

Figure 4.3 The tree, the unicyclic graph, and the bipartite graph with maximum spectral radius for $n = 12$ and $D = 7$.

with equality if and only if $D = 1$ and G is the complete graph K_n, or $D = 2$ and G is the star $K_{1,n-1}$, the cycle C_5, the Petersen graph, the Hoffman-Singleton graph, or the putative 57-regular Moore graph with 3250 vertices.

Dam and Kooij have proved the special case of this theorem for $D = 2$ previously in [154].

Proof. Let u be an arbitrary vertex of G and v be another vertex of G. Let $u = u_0, u_1, \ldots, u_l = v$ be the shortest path between u and v in G. If $l = D$, then $u_0, \ldots, u_l$ is a walk of length D, while if $l < D$, then there exists a walk of length D starting with $u_0, \ldots, u_{l-1}, u_l, u_{l-1}$.

Clearly, each choice $v \neq u$ gives a different walk of length D, which shows that there exists a least $n - 1$ walks of length D starting at u. The total number of walks of length D starting at u is equal to the uth rowsum $(A^D j)_u$, where A is the adjacency matrix of G and j is the all-one vector. Hence,

$$j^T A^D j \geq n(n-1),$$

and the Rayleigh quotient yields

$$\lambda_1^D \geq \frac{j^T A^D j}{j^T j} \geq n - 1. \tag{4.11}$$

Equality is possible only if $D \leq 2$. If $D \geq 3$, then the above count for $l = 1$ of a walk starting with u, v, u can be replaced by $\deg_u^2$ walks u, v, u, v', where v' is also a neighbor of u (the case $v' = v$ is allowed). Since the graph is connected, and not all vertices can have degree 1 due to $D \geq 3$, equality cannot hold in (4.10) for $D \geq 3$.

For $D = 1$, equality holds trivially, as the complete graph K_n is the only graph in this case.

If equality holds for $D = 2$, from (4.11) we see that j must be the eigenvector of A^2 corresponding to its spectral radius λ_1^2. If j is also an eigenvector of A corresponding to its spectral radius λ_1, G has to be r-regular graph for some integer r. Then from $r = \lambda_1 = \sqrt{n-1}$ we get that G has $n = r^2 + 1$ vertices, showing that G is a Moore graph with diameter 2, and by the Hoffman-Singleton theorem [79], G can only be the cycle C_5, the Petersen graph, the Hoffman-Singleton graph, or the putative 57-regular Moore graph with 3250 vertices.

Assume now that j is not an eigenvector of A. Then A also has the eigenvalue $-\lambda_1$, and j is a linear combination of the eigenvectors of A corresponding to λ_1 and $-\lambda_1$. By Theorem 1.2, G is bipartite, so let $V(G) = V_1 \cup V_2$ be the bipartition of its vertices. If x denotes the principal eigenvector of G, the eigenvector x' of $-\lambda_1$ is obtained from x by changing signs of the components in one part of G, i.e.,

$$x_u' = \begin{cases} x_u, & u \in V_1, \\ -x_u, & u \in V_2. \end{cases}$$

The fact that $j = \alpha x + \beta x'$ implies that

$$x_u = \begin{cases} \frac{1}{\alpha+\beta}, & u \in V_1, \\ \frac{1}{\alpha-\beta}, & u \in V_2, \end{cases}$$

and from the eigenvalue equation we immediately conclude that all vertices in the same part have equal degrees, say r in V_1 and s in V_2. Further, it is clear that no two vertices from the same V_i may have two common neighbors (as then we would have more walks of length 2 than accounted for), and, on the other hand, any two vertices from the same V_i have at least one common neighbor in order to keep diameter equal to two, concluding that any 2 vertices from the same V_i have exactly one common neighbor.

This now leads us to balanced incomplete block designs (BIBDs) and the Fisher's inequality. Recall that the BIBD consists of a set of points X

and a family $\mathcal{F}$ of k-element subsets of X, called blocks, such that each point from X is contained in exactly p blocks, and that each pair of distinct points from X is contained in exactly q blocks. The Fisher's inequality [62] states that

$$\text{if } k < |X| \text{ then } |\mathcal{F}| \geq |X|.$$

In our situation, we actually have two BIBDs: in the first one, V_1 is the set of points and the neighborhoods of vertices from V_2, which are s-element subsets of V_1, are its blocks, while in the second one, V_2 is the set of points and the neighborhoods of vertices from V_1 are the r-element blocks. Now, if both $s < |V_1|$ and $r < |V_2|$, then the Fisher's inequality implies that $|V_2| \leq |V_1|$ holds for the first BIBD, and that $|V_1| \leq |V_2|$ holds for the second BIBD, i.e., that $|V_1| = |V_2|$. As the number m of edges in G satisfy $m = |V_1|r = |V_2|s$, this also implies that $r = s$, i.e., that G is regular. However, j is then the eigenvector of A corresponding to λ_1, contrary to our assumption. Hence, it has to be $s = |V_1|$ or $r = |V_2|$ (so that the Fisher's inequality cannot be applied), so that one of these BIBDs contains just one block, in which case G is the star $K_{1,n-1}$. $\square$

Cioabă et al. [36] also posed a conjecture related to the lower bound (4.10). Define

$$\lambda_{n,D}^- = \min\{\lambda_1(G) : G \text{ is a graph with } n \text{ vertices and diameter } D\}.$$

Using Bethe trees and their variations, Cioabă et al. [36] first showed that

$$\limsup_{n \to \infty} \frac{\lambda_{n,D}^-}{\sqrt[D]{n-1}} < 2,$$

and then posed

Conjecture 4.5 ([36]). $\displaystyle \lim_{n \to \infty} \frac{\lambda_{n,D}^-}{\sqrt[D]{n-1}} = 1.$

Moreover, they have shown that this conjecture is implied by the Bollobas' conjecture [19, p. 213] that, for $D \geq 3$,

$$\liminf_{\Delta \to \infty} \frac{n_{\Delta,D}}{\Delta^D} \leq 1,$$

where $n_{\Delta,D}$ is the maximum number of vertices in a graph with the maximum vertex degree Δ and the diameter D.

Belardo et al. [8] have further determined trees with the minimum spectral radius among trees with diameters 3 and 4. Trees with diameter 3 are necessarily double stars, obtained from two stars by joining their centers, and they have shown that the minimum spectral radius is attained by creating a double star out of stars $K_{1,\lfloor n/2-1 \rfloor}$ and $K_{1,\lceil n/2-1 \rceil}$, whose spectral radius is equal to $\sqrt{\frac{n-1}{2} + \sqrt{\frac{n-1}{2}}}$ for odd n, and to $\frac{1}{2}\left(1 + \sqrt{2n-3}\right)$ for even n.

Let us describe construction of the minimum tree with diameter 4. For any $k \geq 0$, let $a_k = k^2 + k + 1$, and for any $n \geq 5$, if $a_k \leq n \leq a_{k+1}$, construct a tree with diameter 4 as follows:

1) $V_0 = \{c\}$, $V_1 = \{v_1, \ldots, v_{k+1}\}$, $V_2 = \{w_1, \ldots, w_{n-k-2}\}$;
2) each vertex from V_1 is adjacent to c;
3) each vertex from V_2 is adjacent to one vertex from V_1, so that $|\deg_{v_i} - \deg_{v_j}| \leq 1$ holds for each $v_i, v_j \in V_1$.

Note that for $n \neq a_{k+1}$ this construction yields a unique tree, denoted by $MT_{n,4}^k$, while for $n = a_{k+1}$ we obtain two non-isomorphic trees $MT_{n,4}^k$ and $MT_{n,4}^{k+1}$, as n then belongs to both intervals $[a_k, a_{k+1}]$ and $[a_{k+1}, a_{k+2}]$.

Theorem 4.14 ([8]). *For any $n \geq 5$, the minimum spectral radius among trees with n vertices and diameter 4 is attained by $MT_{n,4}^k$ if $n \neq k^2 + k + 1$, and by both $MT_{n,4}^k$ and $MT_{n,4}^{k+1}$ if $n = k^2 + k + 1$, for some integer k.*

In cases of graphs with a large diameter, linear in n, the spectral radius decreases towards 2. As a matter of fact, it has been possible to resolve the cases $D \in \{n-1, n-2, n-3, \lfloor n/2 \rfloor\}$ by relying solely on the characterization of graphs with spectral radius at most 2 (the so-called Smith graphs): other than the trivial case of the path P_n for $D = n-1$, van Dam and Kooij [154] have shown that the minimum spectral radius is attained by two Smith graphs D_n and $\tilde{D}_{n-1}$ for $D = n-2$ and $D = n-3$ (see Fig. 4.1), respectively, and the cycle C_n (with a few additional graphs if $n \leq 8$) for $D = \lfloor n/2 \rfloor$.

For other cases, graphs whose spectral radius is at most $\frac{3}{2}\sqrt{2}$ have fundamental importance. Properties of such graphs have been studied by Woo and Neumaier [164], with some further properties obtained by Wang et al. [160]. Following Woo and Neumaier [164], a tree with the maximum vertex degree 3 such that all vertices of degree 3 lie on a path is called an open quipu; a closed quipu is a unicyclic graph with the maximum vertex degree 3 such that all vertices of degree 3 lie on a cycle; and a dagger is obtained from a path by adding three pendant vertices at one of its endvertices. Woo and Neumaier [164] have proved the following

Theorem 4.15 ([164]). *A graph G with $\lambda_1 \leq \frac{3}{2}\sqrt{2}$ (≈ 2.1213) is either an open quipu, a closed quipu, or a dagger.*

Denote by $P^{(m_0,m_1,\ldots,m_r)}_{(k_0,k_1,\ldots,k_r,k_{r+1})}$ the open quipu with r internal paths of lengths $k_1 + 1, \ldots, k_r + 1$ and $r+3$ pendant paths of lengths $k_0, m_0, m_1, \ldots, m_r, k_{r+1}$. Here for $1 \leq i \leq r$, k_i measures the number of internal vertices of the ith internal path of a quipu.

van Dam and Kooij [154] initially posed the following conjecture for graphs with large diameter.

Conjecture 4.6 ([154]). *For fixed e and large enough n, the graph*

$$P^{(\lfloor \frac{e-1}{2} \rfloor, \lceil \frac{e-1}{2} \rceil)}_{(\lfloor \frac{e-1}{2} \rfloor, n-2e, \lceil \frac{e-1}{2} \rceil)}$$

has the minimum spectral radius among graphs with n vertices and the diameter $D = n - e$.

Yuan et al. [166] have proved this conjecture for $D = n - 4$ and $n \geq 11$. Cioabă et al. [37] have given independent proof for $D = n - 4$ and have further proven it for $D = n - 5$ and $n \geq 18$.

However, Cioabă et al. [37] have also shown that Conjecture 4.6 is false for $e \geq n - 6$. They have proved the following

Theorem 4.16 ([37]). *For fixed $e \geq 6$ and large enough n, the graph with the minimum spectral radius among graphs with n vertices and the diameter $D = n - e$ belongs to one of the three families*

$$\mathcal{P}_{n,e} = \{P^{(2,1,\ldots,1,2)}_{(2,k_1,\ldots,k_{e-4},2)} : \textstyle\sum_{i=1}^{e-4} k_i = n - 2e,\ k_i \geq 1\},$$

$$\mathcal{P}'_{n,e} = \{P^{(2,1,\ldots,1,1)}_{(2,k_1,\ldots,k_{e-3},1)} : \textstyle\sum_{i=1}^{e-3} k_i = n - 2e,\ k_i \geq 1\},$$

$$\mathcal{P}''_{n,e} = \{P^{(1,1,\ldots,1,1)}_{(1,k_1,\ldots,k_{e-2},1)} : \textstyle\sum_{i=1}^{e-2} k_i = n - 2e,\ k_i \geq 1\}.$$

This disproves Conjecture 4.6 as $\left\lceil \frac{e-1}{2} \right\rceil \geq 3$ for $e \geq 6$, so that $P^{\left(\left\lfloor \frac{e-1}{2} \right\rfloor, \left\lceil \frac{e-1}{2} \right\rceil\right)}_{\left(\left\lfloor \frac{e-1}{2} \right\rfloor, n-2e, \left\lceil \frac{e-1}{2} \right\rceil\right)}$ does not belong to any of these three families. In turn, Cioabă et al. [37] have posed the following

Conjecture 4.7 ([37]). *For fixed $e \geq 5$ and large enough n, the graph with the minimum spectral radius among graphs with n vertices and the diameter $D = n - e$ belongs to the family $\mathcal{P}_{n,e}$.*

They have been more specific in the cases of $D = n - 6$ and $D = n - 7$.

Conjecture 4.8 ([37]). *For large enough n, the graph $P^{(2,1,2)}_{(2,\lceil(n-13)/2\rceil,\lfloor(n-11)/2\rfloor,2)}$ has the minimum spectral radius among graphs with n vertices and the diameter $D = n - 6$.*

Conjecture 4.9 ([37]). *For large enough n, the graph $P^{(2,1,1,2)}_{(2,\lfloor\frac{n+1}{3}\rfloor-5,n-2\lfloor\frac{n+1}{3}\rfloor-4,\lfloor\frac{n+1}{3}\rfloor-5,2)}$ has the minimum spectral radius among graphs with n vertices and the diameter $D = n - 7$.*

These conjectures have been proved affirmatively by Lan et al. [87].

Theorem 4.17 ([87]). *For any $e \geq 6$ and large enough n, the graph with the minimum spectral radius among graphs with n vertices and the diameter $D = n - e$ is a tree $P^{(2,1,\ldots,1,2)}_{(2,k_1,\ldots,k_{e-4},2)}$ such that*

1) $\lfloor s \rfloor - 1 \leq k_j \leq \lfloor s \rfloor \leq k_i \leq \lceil s \rceil + 1$ for $2 \leq i \leq e - 5$ and $j \in \{1, e - 4\}$, where $s = \frac{n-6}{e-4} - 2$,
2) $0 \leq k_i - k_j \leq 2$ for $2 \leq i \leq e - 5$ and $j \in \{1, e - 4\}$,
3) $|k_i - k_j| \leq 1$ for $2 \leq i,j \leq e - 5$.

In particular, if $n - 6$ is divisible by $e - 4$, then the minimum spectral radius is attained by $P^{(2,1,\ldots,1,2)}_{(2,s-1,s,\ldots,s,s-1,2)}$.

They have also completely determined the extremal graphs in the case $e = 6, 7$, and 8.

Theorem 4.18 ([87]). *For large enough n, the minimum spectral radius among graphs with n vertices and the diameter $D = n - 6$ is attained by*

1) $P^{(2,1,2)}_{(2,k,k,2)}$ if $n = 2k + 12$,

2) $P^{(2,1,2)}_{(2,k,k+1,2)}$ if $n = 2k + 13$.

Theorem 4.19 ([87]). *For large enough n, the minimum spectral radius among graphs with n vertices and the diameter $D = n - 7$ is attained by*

1) $P^{(2,1,1,2)}_{(2,k,k,k,2)}$ if $n = 3k + 14$,

2) $P^{(2,1,1,2)}_{(2,k,k+1,k,2)}$ if $n = 3k + 15$,

3) $P^{(2,1,1,2)}_{(2,k,k+2,k,2)}$ if $n = 3k + 16$.

Theorem 4.20 ([87]). *For large enough n, the minimum spectral radius among graphs with n vertices and the diameter $D = n - 8$ is attained by*

1) $P^{(2,1,1,1,2)}_{(2,k,k,k,k,2)}$, $P^{(2,1,1,1,2)}_{(2,k,k,k+1,k-1,2)}$ *and* $P^{(2,1,1,1,2)}_{(2,k-1,k+1,k+1,k-1,2)}$ if $n = 3k + 16$,

2) $P^{(2,1,1,1,2)}_{(2,k,k+1,k,2)}$ if $n = 3k + 17$,

3) $P^{(2,1,1,1,2)}_{(2,k,k+1,k+1,k,2)}$ if $n = 3k + 18$,

4) $P^{(2,1,1,1,2)}_{(2,k,k+1,k+2,k,2)}$ if $n = 3k + 19$.

Closed quipus take over the role if the diameter D is roughly between $n/2$ and $\frac{2n}{3}$. Denote by $C^{(m_1,...,m_r)}_{(k_1,...,k_r)}$ the closed quipu with r internal paths of lengths $k_1 + 1, \ldots, k_r + 1$ (so, r internal paths with $k_1, \ldots, k_r$ internal vertices of degree 2), and r pendant paths of lengths $m_1, \ldots, m_r$.

Let C^t_s be the family of graphs $C^{(1,...,1)}_{(k_1,...,k_t)}$ such that $\sum_{i=1}^t k_i = s - t$, for some $k_i \geq 0$. Cioabă et al. [37] have proved the following result when $D = (n - e)/2$.

Theorem 4.21 ([37]). *For fixed $e \geq 2$ and large enough n, the graph with the minimum spectral radius among graphs with n vertices and the diameter $D = (n - e)/2$ belongs to one of the four families C^{e+1}_{n-e-1}, C^{e+2}_{n-e-2}, C^{e+3}_{n-e-3}, or C^{e+4}_{n-e-4}.*

For the case $D = (n + e)/2$, Cioabă et al. [37] have posed the following conjecture and proved it for $e \leq 4$.

Conjecture 4.10 ([37]). *For fixed $e \geq 1$ and large enough n such that $n + e$ is even, the graph*

$$C^{(\lfloor\frac{e}{2}\rfloor,\lceil\frac{e}{2}\rceil)}_{(\frac{n-e-2}{2},\frac{n-e-2}{2})}$$

has the minimum spectral radius among graphs with n vertices and the diameter $D = (n + e)/2$.

Lan and Lu [88] have settled this conjecture by proving that the statement holds for all $n \geq 3e + 12$.

Theorem 4.22 ([88]). *For $n \geq 14$ and $\frac{n}{2} \leq D \leq \frac{2n-4}{3}$, the minimum spectral radius among graphs with n vertices and the diameter D is attained by*

1) $C^{(D-\lfloor\frac{n}{2}\rfloor,D-\lceil\frac{n}{2}\rceil)}_{(n-D-1,n-D-1)}$ if $D < \frac{2n-4}{3}$,

2) all $C^{(m,2D-n-m)}_{(n-D-1,n-D-1)}$ for $m \in [0, 2D - n]$ if $D = \frac{2n-4}{3}$.

Transition between closed quipus and open quipus as the graphs with the minimum spectral radius occurs at $D = \frac{2n-3}{3}$. In particular:

Theorem 4.23 ([88]). *For $n \geq 15$ and $D = \frac{2n-3}{3}$, i.e., such that $n = 3k$ and $D = 2k - 1$, the minimum spectral radius among graphs with n vertices and the diameter D is attained by both the closed quipu $C^{(k-2)}_{(2k+1)}$ and the open quipu $P^{(1,k-1)}_{(1,k-2,k-1)}$.*

Lan and Lu [88] have also conjectured the structure of the minimum graph for diameters that are slightly larger than $\frac{2n-3}{3}$.

Conjecture 4.11 ([88]). *For $n \geq 15$ and $D = \frac{2n-2}{3}$, the minimum spectral radius among graphs with n vertices and the diameter D is attained by $P^{(m,m')}_{(m,2D-n,m')}$ for all $m, m' \geq 1$ satisfying $m + m' = n - D - 1$.*

Conjecture 4.12 ([88]). *There exists an positive number ϵ so that for*

$$D \in \left[\frac{2n-1}{3}, \left(\frac{2}{3} + \epsilon + o(1)\right)n\right]$$

and large enough n, the minimum spectral radius among graphs with n vertices and the diameter D is attained by

$$P^{(\lfloor\frac{n-D-1}{2}\rfloor,\lceil\frac{n-D-1}{2}\rceil)}_{(\lfloor\frac{n-D-1}{2}\rfloor,2D-n,\lceil\frac{n-D-1}{2}\rceil)}.$$

To summarize, previous results show that the graph with the minimum spectral radius among graphs with n vertices and the diameter D is known when $D \leq 4$ (assuming that such graphs are trees for $D = 3$ and $D = 4$), when $\frac{n}{2} \leq D \leq \frac{2n-3}{3}$, and when $n - D$ has constant value. However, plenty of work still remains to be done if one returns to the original motivation of van Dam and Kooij [154], to have graphs with both small spectral radius and small diameter. Future work could, therefore, be directed toward graphs with $D = O(\log n)$, which will, in turn, relate this problem more strongly to the well-known degree/diameter problem [102].

4.7 THE RADIUS

The aptly titled manuscript *On bags and bugs* [72] brings yet another novel class of extremal graphs. Let us firstly recall the definition of a bag.

Definition 4.2. A *bag* $Bag_{p,q}$ is a graph obtained from a complete graph K_p by replacing an edge uv with a path P_q. A bag is *odd* if q is odd; otherwise it is *even* (see Fig. 4.4).

The number of vertices in $Bag_{p,q}$ is $n = p + q - 2$ and the number of edges is $m = \binom{p}{2} + q - 2$. As in the previous section, experiments with AutoGraphiX [27] led to a conjecture that an odd bag has the maximum spectral radius among graphs with a given (usual) radius, which was then proved by Hansen and Stevanović in [72]. In the rest of this section, we will closely follow the proof from [72].

Theorem 4.24 ([72]). *Among all graphs with n vertices and radius r, the maximum spectral radius is attained by*

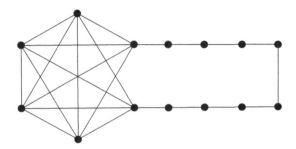

Figure 4.4 A Bag$_{6,10}$.

1) a complete graph K_n when $r = 1$,
2) $\frac{n}{2}\overline{K_2}$ for even n and $r = 2$,
3) $\overline{(\lfloor\frac{n}{2}\rfloor - 1)K_2 \cup P_3}$ for odd n and $r = 2$, and
4) an odd bag $Bag_{n-2r+3,2r-1}$ when $r \geq 3$.

Actually, AutoGraphiX has conjectured that odd bags maximize the spectral radius also in case $r = 2$. However, as we shall see later, odd bags are just the second best candidate for the maximum spectral radius in this case.

Proof. The spectral radius of a graph increases by addition of edges, so for $r = 1$ the maximum spectral radius is attained by a complete graph K_n.

For $r \geq 2$, we have that the maximum spectral radius is attained by a graph G that is radially maximal under addition of new edges: if e is an edge of the complement $\overline{G}$, then $r(G + e) < r(G)$.

For $r = 2$, such graph G is a complement of the union of stars

$$\overline{\bigcup_{i=1}^{m} K_{1,s_i}}, \qquad \sum_{i=1}^{m} s_i = n - m, \qquad m \geq 2, \qquad (4.12)$$

as observed already in [74]. Moreover, G cannot contain vertices of degree $n - 1$, so we have that

$$\lambda_1(G) \leq \Delta \leq n - 2,$$

where Δ is the maximum vertex degree in G. We will determine the candidates for G among the complements of the unions of stars given by (4.12).

Let H be of the form (4.12) and let $\lambda_1(H)$ and x be the spectral radius and the principal eigenvector of H. Consider two stars K_{1,s_i} and K_{1,s_j} in $\overline{H}$. Since $\lambda_1(H)$ is a simple eigenvalue, similar vertices in H, such as the leaves of K_{1,s_i} or the leaves of K_{1,s_j}, have equal components in x. Let a_i be the component of x at the center of K_{1,s_i}, and let b_i be the component of x at any leaf of K_{1,s_i}. Define a_j and b_j similarly. In order to get a relationship between a_i and b_i, consider the eigenvalue equations for the center and a leaf of K_{1,s_i} in H:

$$\lambda_1(H)a_i = \sum_{u \in V(H)\setminus V(K_{1,s_i})} x_u,$$

$$\lambda_1(H)b_i = (s_i - 1)b_i + \sum_{u \in V(H) \setminus V(K_{1,s_i})} x_u,$$

from where it follows that

$$b_i = \frac{\lambda_1(H)a_i}{\lambda_1(H) - s_i + 1}. \tag{4.13}$$

Similarly, we get that

$$b_j = \frac{\lambda_1(H)a_j}{\lambda_1(H) - s_j + 1}. \tag{4.14}$$

Next, in order to get a relationship between a_i and a_j, consider the eigenvalue equations for the centers of K_{1,s_i} and K_{1,s_j} in H:

$$\lambda_1(H)a_i = a_j + s_j b_j + \sum_{u \in V(H) \setminus V(K_{1,s_i} \cup K_{1,s_j})} x_u,$$

$$\lambda_1(H)a_j = a_i + s_i b_i + \sum_{u \in V(H) \setminus V(K_{1,s_i} \cup K_{1,s_j})} x_u,$$

from where it follows that

$$(\lambda_1(H) + 1)\, a_i + s_i b_i = (\lambda_1(H) + 1)\, a_j + s_j b_j.$$

From (4.13) and (4.14), after routine algebraic manipulations, we get that

$$\frac{a_i}{a_j} = \frac{\lambda_1(H) + \frac{\lambda_1(H)+1}{\lambda_1(H)+1-s_j}}{\lambda_1(H) + \frac{\lambda_1(H)+1}{\lambda_1(H)+1-s_i}}.$$

From here we conclude that

$$a_i \le a_j \qquad \Leftrightarrow \qquad s_i \ge s_j.$$

Suppose, without loss of generality, that $s_i \ge s_j \ge 2$. Let H' be a graph obtained by replacing stars K_{1,s_i} and K_{1,s_j} in $\overline{H}$ with stars K_{1,s_i+1} and K_{1,s_j-1}. Graph H' may be obtained from H by rotating an edge between a leaf of K_{1,s_j} and a center of K_{1,s_i} into an edge between a leaf of K_{1,s_j} and a center of K_{1,s_j}. Since $a_i \le a_j$, we have that

$$\lambda_1(H') > \lambda_1(H).$$

We can continue increasing the spectral radius in this way as long as there are at least two stars having three or more vertices. Thus, the candidates for the maximum spectral radius among graphs with n vertices and radius 2 are of the form

$$\overline{K_{1,n-2s-1}} \cup sK_2, \qquad 1 \le s \le \frac{n-3}{2}.$$

Now we have to answer which of these candidates has the largest spectral radius. Let $H_s = \overline{K_{1,n-2s-1}} \cup sK_2$, $1 \le s \le \frac{n-3}{2}$. Similar vertices in H_s, such as the leaves of $K_{1,n-2s-1}$ or the vertices of copies of K_2, have equal components in x, so let a be the component of x at the center of $K_{1,n-2s-1}$, b be the component of x at any leaf of $K_{1,n-2s-1}$, and c the component of x at any vertex of a copy of K_2 in H_s. The eigenvalue equations give

$$\lambda_1(H_s)a = 2sc,$$
$$\lambda_1(H_s)b = (n - 2s - 2)b + 2sc,$$
$$\lambda_1(H_s)c = a + (n - 2s - 1)b + (2s - 2)c.$$

Taking into account that $a, b, c \ne 0$, as the components of the principal eigenvector are positive, we easily get that $\lambda_1(H_s)$ is the largest root of the following polynomial

$$P_s(\lambda) = \lambda^3 - \lambda^2(n - 4) - \lambda(2n - 4) + 2s(n - 2s - 2).$$

(The other two roots are also eigenvalues of H_s.)

It is easy to see that if $P_s(\lambda) < P_t(\lambda)$ for all $\lambda \in \mathbb{R}$ and $P_s(\lambda), P_t(\lambda) \mapsto +\infty$ when $\lambda \mapsto +\infty$, then the largest root of $P_s(\lambda)$ is greater than the largest root of $P_t(\lambda)$. Thus, the maximum spectral radius among the above candidate graphs is obtained when the product $2s(n - 2s - 2)$, $1 \le s \le \lfloor \frac{n-2}{2} \rfloor$, has minimum value. This minimum value is obtained when $s = \lfloor \frac{n-2}{2} \rfloor$: in case n is even, this value is 0 and the maximum spectral radius is attained by $\frac{n}{2}K_2$; in case n is odd, this value is $n - 3$ and the maximum spectral radius is attained by $P_3 \cup \frac{n-3}{2}K_2$. AutoGraphiX [27] has made a mistake here by assuming that the minimum value was obtained for $s = 1$. However, this is indeed the second best choice for s, as other values of s give larger values of product $2s(n - 2s - 2)$.

Now, let $r \ge 3$. An odd bag $Bag_{n-2r+3,2r-1}$ contains the graph $K_{n-2r+3} - e$ as an induced subgraph which has

$$\lambda_1(K_{n-2r+3} - e) = \frac{1}{2}\left(n - 2r + \sqrt{(n - 2r + 3)^2 + 2(n - 2r + 3) - 7}\right),$$

which is strictly larger than $n - 2r + 1$, and so

$$\lambda_1(Bag_{n-2r+3,2r-1}) > n - 2r + 1.$$

Thus G, as a graph with the maximum spectral radius, must have spectral radius also larger than $n - 2r + 1$, and from $\Delta \geq \lambda_1(G)$ we conclude that

$$\Delta \geq n - 2r + 2. \tag{4.15}$$

In order to bound Δ from below, we apply an idea from an old and hard-to-find reference [159]. Let u be a vertex of G of the largest degree Δ, and let $u_1, u_2, \ldots, u_\Delta$ be its neighbors. Let T be a spanning tree of G containing edges $(u, u_1), (u, u_2), \ldots, (u, u_\Delta)$. The radius r_T of T is at least r, and the diameter D_T of T is such that

$$r_T = \left\lfloor \frac{D_T + 1}{2} \right\rfloor.$$

Thus,

$$D_T \geq 2r - 1.$$

Let P be a simple path connecting two most distant vertices of T. Then P contains at most three vertices among $u, u_1, u_2, \ldots, u_\Delta$, and at least $2r - 3$ other vertices. Therefore,

$$\Delta \leq n - 1 - (2r - 3) = n - 2r + 2. \tag{4.16}$$

From (4.15) and (4.16) we conclude that

$$\Delta = n - 2r + 2.$$

If $n = 2r$, then $G \cong C_{2r}$, which is an odd bag $Bag_{3,2r-1}$.

Therefore, suppose that $n \geq 2r + 1$. In particular, we have that

$$\lambda_1(G) > 2, \tag{4.17}$$

a fact that will be used later.

From the above considerations it follows that we must have that $d_T = 2r - 1$, $r_T = r$ and the path P contains all vertices of $V \setminus \{u, u_1, u_2, \ldots, u_\Delta\}$. Denote the vertices of P by $v_1, v_2, \ldots, v_{2r}$. Without loss of generality, we may suppose that path P contains vertices u_1, u, and u_Δ as v_{k-1}, v_k, and v_{k+1}, respectively, for some $2 \leq k \leq 2r - 1$.

Next, let e be an edge in $V(G) \setminus V(T)$. We have that

$$r = r_G \leq r_{T+e} \leq r_T = r,$$

and so G may not contain any edge e, not in T, which reduces the radius of T. From this condition, we easily get that G may not contain edges of the following forms:

1) $v_i v_j$, with $|i - j| > 1$, except for $\{i, j\} = \{1, 2r\}$;
2) $v_i u_j$, with $|i - k| \geq 2$ and $2 \leq j \leq \Delta - 1$;

Moreover, if G contains an edge $u_j v_{k-2}$, $2 \leq j \leq \Delta - 1$, then it may not contain the edges $u_j v_{k+1}$ and $u_j v_{k+2}$, as the presence of any of them would reduce the radius. A similar conclusion holds if G contains an edge $u_j v_{k+2}$. Further, if G contains the edges $u_{j'} v_{k-2}$ and $u_{j''} v_{k+2}$, for some $j' \neq j''$, then it may not contain the edge $u_{j'} u_{j''}$.

Let $L \cup C \cup R$ be the following partition of the set $\{u_2, \ldots, u_\Delta\}$:

$$
\begin{aligned}
L &= \{u_j \mid u_j v_{k-2} \in E\}, \\
C &= \{u_j \mid u_j v_{k-2} \notin E, u_j v_{k+2} \notin E\}, \\
R &= \{u_j \mid u_j v_{k+2} \in E\}.
\end{aligned}
$$

Let $G_{L,C,R} = (V, E_{L,C,R})$ be the graph formed by the following set of edges that may be present in G under the above restrictions:

$$
\begin{aligned}
E_{L,C,R} = {} & \{v_i v_{i+1} \mid i = 1, 2, \ldots, 2r - 1\} \cup \{v_1 v_{2r}\} \\
& \cup \ \{u_j v_{k-2}, u_j v_{k-1} \mid u_j \in L\} \\
& \cup \ \{u_j v_{k-1}, u_j v_{k+1} \mid u_j \in C\} \\
& \cup \ \{u_j v_{k+1}, u_j v_{k+2} \mid u_j \in R\} \\
& \cup \ \{u_i u_j \mid i, j \in L \text{ or } i, j \in C \text{ or } i, j \in R\} \\
& \cup \ \{u_i u_j \mid i \in C, j \in L \cup R\}.
\end{aligned}
$$

An example of a graph $G_{L,C,R}$ is shown in Figure 4.5.

The graph $G_{L,C,R}$ is radially maximal of radius r and, thus, $G \cong G_{L,C,R}$. Our task now is to find the partition $L \cup C \cup R$ yielding a graph $G_{L,C,R}$ with the largest spectral radius. In the rest of the proof we will show that the partition we are looking for is

$$\{L, C, R\} = \{\emptyset, \{u_2, \ldots, u_\Delta\}, \emptyset\},$$

for which $G_{L,C,R}$ is an odd bag.

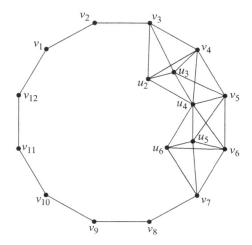

Figure 4.5 A graph $G_{\{u_2,u_3\},\{u_4\},\{u_5,u_6\}}$ [72].

Suppose first that $R = \varnothing$. Suppose also that $L, C \neq \varnothing$. Moving a vertex u_j from L to C represents a rotation around u_j of the edge $u_j v_{k-2}$ onto $u_j v_{k+1}$. If $x_{v_{k+1}} \geq x_{v_{k-2}}$, then moving a vertex u_j from L to C increases the spectral radius, and if $x_{v_{k+1}} < x_{v_{k-2}}$, then moving a vertex u_j from C to L increases the spectral radius. Since G has the maximum spectral radius, this is a contradiction, and so one of L and C must be empty, which finishes the proof in this case.

Next, suppose that $L, R \neq \varnothing$. If $x_{v_{k-2}} \leq x_{v_{k+1}}$, then moving a vertex u_j from L to C represents a rotation around u_j of the edge $u_j v_{k-2}$ onto $u_j v_{k+1}$, followed by addition of edges $u_j u_r$ for $r \in R$. Since each of these operations increases the spectral radius, we have a contradiction to the maximality of G.

Thus, it must hold that

$$x_{v_{k-2}} > x_{v_{k+1}}. \tag{4.18}$$

Similarly, it must hold that

$$x_{v_{k+2}} > x_{v_{k-1}}. \tag{4.19}$$

Next, we show that this also leads to a contradiction.

In the sequel, we will use operations module $2r$ with results in $\{1, 2, \ldots, 2r\}$ when referring to the indices of the vertices $v_1, \ldots, v_{2r}$. From the eigenvalue equation and (4.17) we have that

$$2x_{v_{k+r}} < \lambda_1(G)x_{v_{k+r}} = x_{v_{k+r-1}} + x_{v_{k+r+1}},$$

which is the induction basis. Suppose that we have already proved that

$$x_{v_{k+r-(i-1)}} + x_{v_{k+r+(i-1)}} < x_{v_{k+r-i}} + x_{v_{k+r+i}} \tag{4.20}$$

for some $i \geq 1$. Next, if v_{k+r-i} and v_{k+r+i} are of degree 2, then

$$2(x_{v_{k+r-i}} + x_{v_{k+r+i}}) < \lambda_1(G)(x_{v_{k+r-i}} + x_{v_{k+r+i}})$$
$$= (x_{v_{k+r-(i-1)}} + x_{v_{k+r+(i-1)}}) + (x_{v_{k+r-(i+1)}} + x_{v_{k+r+(i+1)}}).$$

From the last inequality and (4.20) it follows that

$$x_{v_{k+r-i}} + x_{v_{k+r+i}} < x_{v_{k+r-(i+1)}} + x_{v_{k+r+(i+1)}},$$

which proves the inductive step. In particular, we conclude that

$$x_{v_{k-3}} + x_{v_{k+3}} < x_{v_{k-2}} + x_{v_{k+2}}.$$

We cannot apply induction further, as v_{k-2} and v_{k+2} are adjacent to vertices from L and R, respectively, and so have degree larger than 2.

From (4.18) and (4.19) we have that

$$x_{v_{k-1}} + x_{v_{k+1}} < x_{v_{k-2}} + x_{v_{k+2}}. \tag{4.21}$$

The eigenvalue equations for v_{k-2} and v_{k+2} give

$$\lambda_1(G)(x_{v_{k-2}} + x_{v_{k+2}}) = (x_{v_{k-1}} + x_{v_{k+1}}) + (x_{v_{k-3}} + x_{v_{k+3}}) + \sum_{u_j \in L} x_{u_j} + \sum_{u_j \in R} x_{u_j}$$
$$< 2(x_{v_{k-2}} + x_{v_{k+2}}) + \sum_{u_j \in L} x_{u_j} + \sum_{u_j \in R} x_{u_j}.$$

Therefore,

$$(\lambda_1(G) - 2)(x_{v_{k-2}} + x_{v_{k+2}}) < \sum_{u_j \in L} x_{u_j} + \sum_{u_j \in R} x_{u_j}. \tag{4.22}$$

Next, (4.22) and the eigenvalue equations for v_{k-1} and v_{k+1} give

$$\lambda_1(G)(x_{v_{k-1}} + x_{v_{k+1}}) \tag{4.23}$$
$$= 2x_{v_k} + (x_{v_{k-2}} + x_{v_{k+2}}) + 2\sum_{u_j \in C} x_{u_j} + \sum_{u_j \in L} x_{u_j} + \sum_{u_j \in R} x_{u_j}$$
$$> 2x_{v_k} + (x_{v_{k-2}} + x_{v_{k+2}}) + (\lambda_1(G) - 2)(x_{v_{k-2}} + x_{v_{k+2}})$$
$$= 2x_{v_k} + (\lambda_1(G) - 1)(x_{v_{k-2}} + x_{v_{k+2}}).$$

Combined with (4.21) we get that

$$x_{v_{k-2}} + x_{v_{k+2}} > 2x_{v_k}. \tag{4.24}$$

Finally, (4.23), (4.22), and the eigenvalue equation for x_{v_k} gives

$$\begin{aligned}
\lambda_1(G)x_{v_k} &= (x_{v_{k-1}} + x_{v_{k+1}}) + \sum_{u_j \in C} x_{u_j} + \sum_{u_j \in L} x_{u_j} + \sum_{u_j \in R} x_{u_j} \\
&> \frac{2}{\lambda_1(G)}x_{v_k} + \frac{\lambda_1(G) - 1}{\lambda_1(G)}(x_{v_{k-2}} + x_{v_{k+2}}) \\
&+ (\lambda_1(G) - 2)(x_{v_{k-2}} + x_{v_{k+2}}),
\end{aligned}$$

from which it follows that

$$(\lambda_1^2(G) - 2)x_{v_k} > (\lambda_1^2(G) - \lambda_1(G) - 1)(x_{v_{k-2}} + x_{v_{k+2}}).$$

Together with (4.24) we get that

$$x_{v_{k-2}} + x_{v_{k+2}} > 2 \cdot \frac{\lambda_1^2(G) - \lambda_1(G) - 1}{\lambda_1^2(G) - 2} \cdot (x_{v_{k-2}} + x_{v_{k+2}}).$$

Since x is positive, it follows that

$$1 > 2 \cdot \frac{\lambda_1^2(G) - \lambda_1(G) - 1}{\lambda_1^2(G) - 2}.$$

The solution of this quadratic inequality is

$$\lambda_1(G) \in (-\sqrt{2}, 0) \cup (\sqrt{2}, 2),$$

which is in contradiction with (4.17). □

4.8 THE DOMINATION NUMBER

We are interested here in graphs with the maximum spectral radius among the graphs with a given value of domination number γ. In doing this, we will closely follow [143]. We will deal with two sets of graphs separately: all graphs with domination number γ, and graphs with domination number γ having no isolated vertices. The examples of graphs with no isolated vertices and a given domination number were found using AutoGraphiX, which helped to formulate the structure of extremal graphs.

Definition 4.3 ([143]). The *surjective split graph* $SSG(n, k; a_1, \ldots, a_k)$, defined for positive integers $n, k, a_1, \ldots, a_k, n \geq k \geq 3$, satisfying $a_1 + \cdots +$

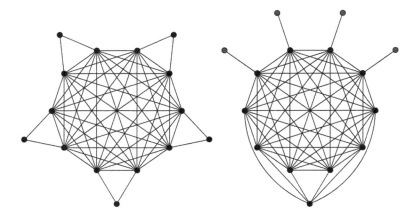

Figure 4.6 The surjective split graphs SSG(15, 5; 2, 2, 2, 2, 2) and SSG(15, 5; 6, 1, 1, 1, 1) [143].

$a_k = n - k$, $a_1 \geq \cdots \geq a_k$, is a split graph on n vertices formed from a clique K with $n - k$ vertices and an independent set I with k vertices, in such a way that the ith vertex of I is adjacent to a_i vertices of K, and that no two vertices of I have a common neighbor in K (see Fig. 4.6).

Note that

$$\gamma(SSG(n, k; a_1, \ldots, a_k)) = k.$$

It is known (see [129]) that surjective split graphs have maximum number of edges among graphs with no isolated vertices and a given domination number $\gamma \geq 3$.

Before we move on to show which surjective split graphs have the maximum spectral radius, we need two lemmas. The first one provides a rather tight, the second-order estimate on the spectral radius of $SSG(n, \gamma; n - 2\gamma + 1, 1, 1, \ldots, 1)$ (note that $\gamma \leq \frac{n}{2}$ for graphs with no isolated vertices [118]).

Lemma 4.5 ([143]). *For the spectral radius of* $SSG(n, \gamma; n - 2\gamma + 1, 1, 1, \ldots, 1)$ *holds*

$$\lambda_1(SSG(n, \gamma; n - 2\gamma + 1, 1, 1, \ldots, 1)) \geq n - \gamma - 1 + \frac{1}{n - \gamma}$$
$$+ \frac{(n - 2\gamma + 1)(n - 2\gamma)}{(n - \gamma)^2},$$

with equality if and only if $\gamma = 1$.

Proof. Let $S^* = SSG(n, \gamma; n-2\gamma+1, 1, 1, \ldots, 1)$. Let η denote the value on the right-hand side of the above inequality. It is easy to see that

$$\eta = n - \gamma - \frac{(2n - 3\gamma)(\gamma - 1)}{(n - \gamma)^2} \le n - \gamma.$$

Let $I' = \{u_a\}$ be the subset of the independent set I of S^* containing the vertex of degree $n - 2\gamma + 1$, and let K' be the subset of the clique K of S^* containing the $n - 2\gamma + 1$ vertices adjacent to u_a. Set $I'' = I \setminus I'$ and $K'' = K \setminus K'$. From the definition of S^*, each vertex of K'' is adjacent to a unique vertex of I'', and $|K''| = |I''| = \gamma - 1$.

Now, let $y = (y_u)_{u \in V(S^*)}$ be the vector defined in the following way:

$$y_u = \begin{cases} a = (n - 2\gamma + 1)\left(1 + \frac{n-2\gamma}{(n-\gamma)^2}\right), & u \in I', \\ b = \eta\left(1 + \frac{n-2\gamma}{(n-\gamma)^2}\right), & u \in K', \\ c = \eta, & u \in K'', \\ d = 1, & u \in I''. \end{cases}$$

For $A = A(S^*)$, we have

$$(Ay)_u = \begin{cases} (n - 2\gamma + 1)b, & u \in I', \\ a + (n - 2\gamma)b + (\gamma - 1)c, & u \in K', \\ (n - 2\gamma + 1)b + (\gamma - 2)c + d, & u \in K'', \\ c, & u \in I''. \end{cases}$$

Let us show that for this particular vector the inequality

$$Ay \ge \eta y,$$

holds componentwise. Actually, for $u \in I$ we have equality:

$$(Ay)_u = \eta(n - 2\gamma + 1)\left(1 + \frac{n-2\gamma}{(n-\gamma)^2}\right) = \eta y_u, \quad u \in I',$$
$$(Ay)_u = \eta = \eta y_u, \quad u \in I''.$$

Next, for $u \in K'$ we have

$(Ay)_u$
$$= n - 2\gamma + 1 + \eta(n - \gamma - 1) + \frac{n - 2\gamma}{(n - \gamma)^2}(n - 2\gamma + 1 + (n - 2\gamma)\eta)$$
$$\ge n - 2\gamma + \eta\left(\frac{1}{n - \gamma} + n - \gamma - 1\right) + \frac{n - 2\gamma}{(n - \gamma)^2}(n - 2\gamma + 1 + (n - 2\gamma)\eta)$$

$$= \eta \left(\eta - \frac{(n - 2\gamma + 1)(n - 2\gamma)}{(n - \gamma)^2} \right)$$

$$+ \frac{n - 2\gamma}{(n - \gamma)^2} \left((n - \gamma)^2 + n - 2\gamma + 1 + (n - 2\gamma)\eta \right)$$

$$= \eta^2 + \frac{n - 2\gamma}{(n - \gamma)^2} \left((n - \gamma)^2 + n - 2\gamma + 1 - \eta \right)$$

$$> \eta^2 \left(1 + \frac{n - 2\gamma}{(n - \gamma)^2} \right)$$

$$= \eta y_u,$$

where in the first inequality above we used the relation $1 \geq \frac{\eta}{n-\gamma}$, and the second inequality, based on $(n - \gamma)^2 + n - 2\gamma + 1 - \eta > \eta^2$, follows from

$$(n - \gamma)^2 - \eta^2 + n - \gamma - \eta = (n - \gamma - \eta)(n - \gamma + \eta + 1)$$

$$> \frac{(2n - 3\gamma)(\gamma - 1)}{(n - \gamma)^2} \cdot 2(n - \gamma)$$

$$> \gamma - 1,$$

thanks to the fact that $\eta > n - \gamma - 1$ and $2(2n - 3\gamma) > n - \gamma$.

Finally, for $u \in K''$ we have

$$(Ay)_u = (n - 2\gamma + 1)\eta \left(1 + \frac{n - 2\gamma}{(n - \gamma)^2} \right) + (\gamma - 2)\eta + 1$$

$$= \eta \left(n - \gamma - 1 + \frac{(n - 2\gamma + 1)(n - 2\gamma)}{(n - \gamma)^2} \right) + 1$$

$$= \eta \left(\eta - \frac{1}{n - \gamma} \right) + 1 = \eta^2 - \frac{\eta}{n - \gamma} + 1$$

$$\geq \eta^2 = \eta y_u.$$

Therefore, we see that

$$\lambda_1(S^*) = \sup_{x \neq 0} \frac{x^T A x}{x^T x} \geq \frac{y^T A y}{y^T y} \geq \frac{y^T (\eta y)}{y^T y} = \eta.$$

□

The next lemma is of independent interest and implies the existence of a number of vertices of maximum vertex degree Δ in a graph whose spectral radius is sufficiently close to $\Delta - 1$.

Lemma 4.6 ([143]). *If a graph G with the maximum vertex degree Δ satisfies*

$$\lambda_1 \geq \Delta - 1 + \frac{k}{\Delta}, \qquad 1 \leq k < \Delta,$$

then G contains at least $k + 1$ vertices of degree Δ.

Proof. Suppose the contrary, i.e., that G contains l vertices with degree Δ, $l \leq k$, while the degree of any other vertex is at most $\Delta - 1$. Let d_v denotes the degree of a vertex $v \in V(G)$. Define a vector $y = (y_v)_{v \in V(G)}$ by

$$y_v = \begin{cases} 1 + \frac{1}{\Delta}, & \text{if } d_v = \Delta, \\ 1, & \text{if } d_v \leq \Delta - 1. \end{cases}$$

For the adjacency matrix $A = A(G)$ we have that, when $d_v = \Delta$,

$$(Ay)_v \leq \Delta + \frac{l-1}{\Delta} < \Delta + \frac{k-1}{\Delta} + \frac{k}{\Delta^2} = \left(\Delta - 1 + \frac{k}{\Delta}\right) y_v,$$

while when $d_v \leq \Delta - 1$,

$$(Ay)_v \leq d_v + \frac{k}{\Delta} \leq \Delta - 1 + \frac{k}{\Delta} = \left(\Delta - 1 + \frac{k}{\Delta}\right) y_v.$$

Thus, for the positive vector y the inequality

$$Ay \leq \left(\Delta - 1 + \frac{k}{\Delta}\right) y$$

holds componentwise.

Let x be the principal eigenvector of G. Then $x^T y > 0$ and we have

$$\lambda_1 x^T y = x^T Ay \leq \left(\Delta - 1 + \frac{k}{\Delta}\right) x^T y,$$

from where it follows that $\lambda_1 \leq \Delta - 1 + \frac{k}{\Delta}$. Note that λ_1, as an algebraic integer, is either an integer or an irrational number. Hence, it cannot not happen that $\lambda_1 = \Delta - 1 + \frac{k}{\Delta}$. However, the conclusion $\lambda_1 < \Delta - 1 + \frac{k}{\Delta}$ is a contradiction with theorem's assertion, so that we conclude that G contains at least $k + 1$ vertices of degree Δ. $\square$

Remark 4.1. In general, we may not get a stronger conclusion from the hypothesis $\lambda_1 > \Delta - 1 + \frac{k}{\Delta}$. Namely, provided that $n(\Delta - 1)$ is even and $\Delta > \frac{n}{2}$, for $k = 1$, there exist graphs with $\lambda_1 > \Delta - 1 + \frac{1}{\Delta}$ which have exactly two vertices of degree Δ. For example, let G be obtained from an arbitrary $(\Delta - 1)$-regular graph on n vertices by joining two of its nonadjacent vertices by an edge. Then the average vertex degree of G is $\Delta - 1 + \frac{2}{n}$. The spectral radius λ_1 is bounded from below by the average vertex degree of a graph [43], so we have $\lambda_1 \geq \Delta - 1 + \frac{2}{n} > \Delta - 1 + \frac{1}{\Delta}$.

We are now ready for the main results of this section. The first theorem deals with the easier case when the graph is allowed to contain isolated vertices (which, understandably, cannot be dominated by any other vertex but itself, and thus, allow easy control of the domination number).

Theorem 4.25 ([143]). *If G is a graph on n vertices with domination number γ, then*

$$\lambda_1 \leq n - \gamma.$$

Equality holds if and only if $G \cong K_{n-\gamma+1} \cup (\gamma - 1)K_1$ or, when $n - \gamma$ is even, $G \cong \frac{n-\gamma+2}{2}K_2 \cup (\gamma - 2)K_1$.

Proof. Let G be a graph with domination number γ, and let G_1, $G_2, \ldots, G_t$ be its connected components. Then $\lambda_1 = \max_{1 \leq i \leq t} \lambda_1(G_i)$. The maximum vertex degree $\Delta(G)$ of G satisfies

$$\Delta(G) \leq n - \gamma, \tag{4.25}$$

since if a vertex u has more than $n - \gamma$ neighbors, then u and its nonneighbors form a dominating set with fewer than γ vertices, a contradiction. Next, from $\lambda_1 \leq \Delta(G)$ it easily follows that

$$\lambda_1 \leq n - \gamma. \tag{4.26}$$

The rest of the proof deals with the case of equality. Suppose that equality holds in (4.26). Then we have $\lambda_1 = \Delta(G) = n - \gamma$. The first of these equalities holds if and only if one of the components containing a vertex of degree $\Delta(G)$ say G_1, is $\Delta(G)$-regular, while for the other components, $2 \leq i \leq t$, it holds that $\lambda_1(G_i) \leq \Delta(G)$.

Let u be a vertex of G_1. It has $n - \gamma$ neighbors (in G_1) dominated by u and $\gamma - 1$ nonneighbors (which include $G_2, \ldots, G_t$) which dominate themselves. If any two nonneighbors v, w of u are adjacent, then we can form a dominating set in G of size less than γ by choosing u, one of v, w only and then the remaining nonneighbors of u. Thus, the nonneighbors of u

must not be adjacent. In particular, the components $G_2, \ldots, G_t$ consist of isolated vertices. As a consequence, we get that G_1 has $n - t + 1$ vertices.

If u is adjacent to all vertices of G_1, then G_1 as a $(n - \gamma)$-regular graph on $n - \gamma + 1$ vertices must be isomorphic to $K_{n-\gamma+1}$, and so we have $G \cong K_{n-\gamma+1} \cup (\gamma - 1)K_1$.

If u has one nonneighbor in G_1, then G_1 as a $(n - \gamma)$-regular graph on $n - \gamma + 2$ vertices must be isomorphic to $\frac{n-\gamma+2}{2}K_2$, and so we have $G \cong \frac{n-\gamma+2}{2}K_2 \cup (\gamma - 2)K_1$.

Thus, suppose that u has at least two nonneighbors in G_1. If any two nonneighbors v, w of u have a common neighbor s in G_1, then we can again form a dominating set of G of size less than γ by taking u, s, and the remaining nonneighbors of u. Therefore, no two nonneighbors of u can have a common neighbor in G_1. This implies that the closed neighborhoods of nonneighbors of u are mutually disjoint. Each of these $\gamma - t$ closed neighborhoods contains $n - \gamma + 1$ vertices, while none of them contains u. Thus, we have the following inequality:

$$(\gamma - t)(n - \gamma + 1) \le n - t,$$

from where it follows that

$$(\gamma - t - 1)n \le (\gamma - 1)(\gamma - t) - t,$$

and so,

$$n \le \frac{(\gamma - t)(\gamma - 1) - t}{\gamma - t - 1} = \frac{(\gamma - t - 1)(\gamma - 1) + \gamma - 1 - t}{\gamma - t - 1} = \gamma \le n.$$

Therefore, $n = \gamma$. In such case, G consists of isolated vertices only, i.e., $G_1 \cong K_1$, and we have a contradiction as u does not have two nonneighbors in G_1.

At last, it is trivial to check that $K_{n-\gamma+1} \cup (\gamma - 1)K_1$ and, for $n - \gamma$ even, $\frac{n-\gamma+2}{2}K_2 \cup (\gamma - 2)K_1$ have spectral radius $n - \gamma$ and domination number γ. $\square$

The next theorem deals with the much more interesting case of graphs with no isolated vertices.

Theorem 4.26 ([143]). *If G is a graph on n vertices with no isolated vertices and domination number γ, then for its spectral radius λ_1 holds*

1) if $\gamma = 1$, then $\lambda_1 \leq \lambda_1(K_n)$, with equality if and only if $G \cong K_n$;

2) if $\gamma = 2$ and n is even, then $\lambda_1 \leq \lambda_1 \left(\frac{n}{2} K_2 \right)$, with equality if and only if $G \cong \overline{\frac{n}{2} K_2}$;

3) if $\gamma = 2$ and n is odd, then $\lambda_1 \leq \lambda_1 \left(\overline{\left(\frac{n}{2} - 1 \right) K_2 \cup P_3} \right)$, with equality if and only if $G \cong \overline{\left(\frac{n}{2} - 1 \right) K_2 \cup P_3}$;

4) if $3 \leq \gamma \leq \frac{n}{2}$, then

$$\lambda_1 \leq \lambda_1(SSG(n, \gamma; n - 2\gamma + 1, 1, 1, \ldots, 1)),$$

with equality if and only if $G \cong SSG(n, \gamma; n - 2\gamma + 1, 1, 1, \ldots, 1)$.

Proof. (1)   Due to the monotonicity of the spectral radius with respect to edge addition, for any graph G on n vertices holds $\lambda_1 \leq \lambda_1(K_n)$, with equality if and only if $G \cong K_n$. Since the complete graph K_n has domination number 1, this completes the case $\gamma = 1$.

(2),(3)   Suppose that $\gamma = 2$ and let G^* be a graph having the maximum spectral radius among all graphs on n vertices with no isolated vertices and domination number 2.

If G^* is disconnected with components $G_1, \ldots, G_t, t \geq 2$, then each component has at least two vertices, and thus, the biggest component has at most $n - 2$ vertices. Consequently, as $\lambda_1(G^*) = \max_{1 \leq i \leq t} \lambda_1(G_i)$, we have $\lambda_1(G^*) \leq \lambda_1(K_{n-2}) = n - 3$.

On the other hand, suppose that G^* is connected. Due to the monotonicity of the spectral radius, we see that G^* has to be *domination critical*: the graph $G^* + e$ has domination number less than γ for every edge e that does not belong to G^*. The structure of the domination-critical graphs with domination number 2 has been determined in [148]: a graph with domination number 2 is domination critical if and only if it is isomorphic to $\overline{\bigcup_{i=1}^{t} K_{1,n_i}}$ for some $n_1, n_2, \ldots, n_t$.

Thus, G^* is a complement of the union of stars, and consequently, it is also a radius-critical graph with radius 2 [74]. Therefore, G^*, which has the maximum spectral radius among all connected graphs on n vertices with domination number 2, also has maximum spectral radius among all connected graphs on n vertices with radius 2. However, from Section 4.7 we already know that in such case

$$G^* \cong \overline{\frac{n}{2}K_2} \text{ for even } n \qquad \text{and} \qquad G^* \cong \overline{\left(\lfloor \frac{n}{2} \rfloor - 1\right) K_2 \cup P_3} \text{ for odd } n.$$

Since both of these graphs have average vertex degree larger than $n - 3$, and since the spectral radius of a graph is larger than or equal to its average vertex degree, we conclude that $\overline{\frac{n}{2}K_2}$ for even n and $\overline{\left(\lfloor \frac{n}{2} \rfloor - 1\right) K_2 \cup P_3}$ for odd n are indeed the graphs which maximize the spectral radius among graphs with no isolated vertices and domination number 2.

(4)   Let $3 \leq \gamma \leq \frac{n}{2}$ and let G^* be a graph on n vertices with no isolated vertices, domination number γ and the maximum spectral radius $\lambda_1^* = \lambda_1(G^*)$. Let S^* be the surjective split graph $SSG(n, \gamma; n - 2\gamma + 1, 1, \ldots, 1)$. The rest of the proof is rather involved, so we list the main idea at the beginning of each new part of the proof.

G^* has at least two vertices of degree $n - \gamma$.

From $\Delta(G^*) \geq \lambda_1^*$, then the fact that S^* has domination number γ and Lemma 4.5, we get

$$\Delta(G^*) \geq \lambda_1^* \geq \lambda_1(S^*) \geq n - \gamma - 1 + \frac{1}{n - \gamma}.$$

Together with (4.25), this yields

$$\Delta(G^*) = n - \gamma.$$

Lemma 4.6 now implies that G^* contains at least two vertices of degree $n - \gamma$. Suppose that w' and w'' are vertices of degree $n - \gamma$.

G^* is connected.

On the contrary, suppose that G^* has components $G_1^*, G_2^*, \ldots, G_t^*$, $t \geq 2$. Suppose that w' belongs to G_1^*. Vertex w' together with its $\gamma - 1$ nonneighbors forms a minimal dominating set D in G^*. There exists no edge between any two nonneighbors of w', as the existence any such edge yields a smaller dominating set. In particular, it follows that the components $G_2^*, \ldots, G_t^*$ consist of isolated vertices. However, this is a contradiction with our premise that G^* has no isolated vertices.

G^* is domination critical.

If any edge e may be added to G^* without decreasing its domination number, then $G^* + e$ has domination number γ and strictly larger spectral radius than G^*, which is a contradiction. Thus, no edge may be added to G^* without decreasing its domination number, and so, G^* is domination critical.

The local structure of G^* imposed by w'.

Let

$$S_{w'} = \{s_1, s_2, \ldots, s_{\gamma-1}\}$$

be the set of $\gamma - 1$ vertices that are not adjacent to w'. The subgraph induced by $S_{w'}$ contains no edges: any edge between vertices of $S_{w'}$ leads to a dominating set of size $\gamma - 1$, a contradiction.

Similarly, no two vertices from $S_{w'}$ may have a common neighbor: for if t is a vertex of G^* adjacent to vertices u and v of $S_{w'}$, then $\{w', t\} \cup S_{w'} \setminus \{u, v\}$ would be again a dominating set of size $\gamma - 1$.

Let $Y_{w'}$ be the set of vertices that are adjacent both to w' and to a vertex from $S_{w'}$. In particular, for each $u \in S_{w'}$, let $Y_{w',u}$ be the set of vertices that are adjacent to w' and u. The set $Y_{w',u}$ is not empty, as G^* does not contain isolated vertices, and from the previous paragraph it follows that each neighbor of u must also be a neighbor of w'. Moreover, it also follows that the sets $Y_{w',u}$, $u \in S_{w'}$, are mutually distinct.

Finally, let $Z_{w'}$ be the set of remaining vertices of G^*, those which are adjacent to w' and to no vertex of $S_{w'}$. The set $Z_{w'}$ is not empty: otherwise, a set X obtained by choosing an arbitrary vertex from each $Y_{w',u}$, $u \in S_{w'}$, would be a dominating set of size $\gamma - 1$. Actually, for each such X an even stronger statement holds:

There exists a vertex z_X in $Z_{w'}$ that is not adjacent to any vertex in X.

(4.27)

We may now see that for any $u \in S_{w'}$, every dominating set X of G^* must contain either the vertex u or a vertex from $Y_{w',u}$. In particular, if $|X| = \gamma$, then $\gamma - 1$ vertices of X belong to the sets $\{u\} \cup Y_{w',u}$, $u \in S_{w'}$, and the remaining vertex belongs to $\{w'\} \cup Z_{w'}$.

Next, the subgraph of G^* induced by $Y_{w'}$ is a clique: otherwise, if uv is not an edge of G^* for $u, v \in Y_{w'}$, then $G^* + uv$ also has domination number γ, but its spectral radius is larger than λ_1^*, a contradiction. From a similar reason, the subgraph induced by $Z_{w'}$ is also a clique.

Where does w'' appear: in $S_{w'}$, $Y_{w'}$, or $Z_{w'}$?

The only part of G^* that we do not know anything about is the set of edges between vertices of $Y_{w'}$ and $Z_{w'}$. This is where the second vertex w'' of degree $n - \gamma$ helps us. Note that the sets $S_{w''}$, $Y_{w''}$, and $Z_{w''}$ may be defined in the same manner and share similar properties as their counterparts $S_{w'}$, $Y_{w'}$, and $Z_{w'}$. So, let us consider in which of the three sets $S_{w'}$, $Y_{w'}$, and $Z_{w'}$ may w'' appear?

The case $w'' \in S_{w'}$ is impossible.

First, w'' may not belong to $S_{w'}$, as the degrees of vertices in $S_{w'}$ are too small. Namely, a vertex $u \in S_{w'}$ is not adjacent to any vertex from

$$\{w'\} \cup Z_{w'} \cup (S_{w'} \setminus \{u\}) \cup (Y_{w'} \setminus Y_{w',u}),$$

and its degree is, thus, at most $n - 1 - (1 + 1 + (\gamma - 2) + (\gamma - 2)) < n - \gamma$.

If $w'' \in Y_{w'}$, then G^* is a surjective split graph.

Next, suppose that $w'' \in Y_{w'}$ and let s be the unique vertex of $S_{w'}$ adjacent to w''. Then w'' is adjacent to all vertices of $Z_{w'}$ but one, which we denote by z. It is easy to see that

$$\begin{aligned} S_{w''} &= \{z\} \cup S_{w'} \setminus \{s\}, \\ Y_{w''} &\supseteq \{w'\} \cup (Y_{w'} \setminus Y_{w',s}) \cup (Z_{w'} \setminus \{z\}), & (4.28) \\ Z_{w''} &\subseteq \{s\} \cup Y_{w',s} \setminus \{w''\}. & (4.29) \end{aligned}$$

We show that equality holds in (4.28) and (4.29). Suppose that $t \in Y_{w',s} \cap Y_{w''}$. Since the subgraph induced by $Y_{w''}$ is a clique, t must be adjacent to all vertices of $Z_{w'} \setminus \{z\}$. Further, as an element of $Y_{w''}$, t must be adjacent to a vertex of $S_{w''}$. Since it is not adjacent to any vertex of $S_{w'} \setminus \{s\}$, we conclude that t is adjacent to z as well. But then t has degree $n - \gamma + 1$, a contradiction. Thus, it follows that $Y_{w',s} \cap Y_{w''} = \emptyset$ and then the equality holds in (4.28) and (4.29). Moreover, one has

$$Y_{w'',u} = Y_{w',u}, \quad u \in S_{w'} \setminus \{s\}$$

and

$$Y_{w'',z} = \{w'\} \cup Z_{w'} \setminus \{z\}.$$

As a consequence, z is adjacent to vertices of $Y_{w'',z}$ only, and so z is not adjacent to any vertex from $Y_{w'}$. Then G^*, as a domination-critical graph, must already contain all edges between a vertex of $Y_{w'}$ and $Z_{w'} \setminus \{z\}$. In such case, G^* is indeed a surjective split graph:

$$G^* \cong SSG(n, \gamma; |Z_{w'}|, |Y_{w',s_1}|, \ldots, |Y_{w',s_{\gamma-1}}|).$$

If $w'' \in Z_{w'}$, then G^* is a surjective split graph.

The final option for w'' is that it belongs to $Z_{w'}$. We may freely suppose then that no vertex of $Y_{w'}$ has degree $n - \gamma$ (otherwise, rename any such vertex to w'' and return to the previous paragraph). Let U be the set of all vertices of G^* having degree $n - \gamma$. Then $U \subseteq \{w'\} \cup Z_{w'}$. The vertices of U imply the same local structure in G^*: for any $w \in U$ one has

$$\begin{aligned} S_w &= S_{w'}, \\ Y_w &= Y_{w'}, \\ Z_w &= \{w'\} \cup Z_{w'} \setminus \{w\}. \end{aligned}$$

Finally, let $Z' = Z_{w'} \setminus U$. Any vertex $z' \in Z'$ has degree less than $n - \gamma$ and, thus, there exists a vertex $y' \in Y_{w'}$ not adjacent to z'. The graph $G^* + y'z'$ has a dominating set X of cardinality $\gamma - 1$. Note that $y' \in X \subseteq Y_{w'}$ and that X does not dominate z' in G^*. Thus, z' is not adjacent to any vertex of X in G^*. In other words, for any $z' \in Z'$, G^* does not contain at least $\gamma - 1$ edges of the form $z'v$. This can be used to give an upper bound on the number of edges m^* of G^*:

$$m^* \le \binom{n - \gamma + 1}{2} - |Z'|(\gamma - 1) + (\gamma - 1). \tag{4.30}$$

(The last term above counts the edges between $S_{w'}$ and $Y_{w'}$.)

We can pair this upper bound with a lower bound on m^* obtained from Lemma 4.5 and the bound of Hong [82]

$$\lambda_1(G^*) \le \sqrt{2m^* - n + 1}.$$

Namely, we have

$$2m^* \ge \lambda_1(G^*)^2 + n - 1$$

$$= \left(n - \gamma - \frac{(2n - 3\gamma)(\gamma - 1)}{(n - \gamma)^2}\right)^2 + n - 1$$

$$\geq (n - \gamma)^2 - \frac{2(2n - 3\gamma)(\gamma - 1)}{n - \gamma} + n - 1$$

$$> (n - \gamma)^2 - 4(\gamma - 1) + n - 1 = (n - \gamma + 1)(n - \gamma) - 3(\gamma - 1),$$

i.e.,

$$m^* > \binom{n - \gamma + 1}{2} - \frac{3}{2}(\gamma - 1). \tag{4.31}$$

Inequalities (4.30) and (4.31), taken together, yield:

$$|Z'| \leq \frac{5}{2}.$$

Note that the case $|Z'| = 2$ is impossible. Namely, since each vertex $y' \in Y_{w'}$ has degree less than $n - \gamma$, there are at least two vertices in Z' not adjacent to y'. Thus, neither of two vertices of Z' is adjacent to any vertex of $Y_{w'}$. However, we can then add to G^* all edges between one vertex of Z' and all vertices of $Y_{w'}$ without decreasing its domination number, which is a contradiction.

Thus, $|Z'| = 1$. Then G^* is again a surjective split graph

$$G^* \cong SSG(n, \gamma; |U|, |Y_{w',s_1}|, \ldots, |Y_{w',s_{\gamma-1}}|).$$

S^* has maximum spectral radius among surjective split graphs.

Finally, we may suppose that $G^* \cong SSG(n, \gamma; a_1, \ldots, a_\gamma)$ for $a_1 \geq \cdots \geq a_\gamma \geq 1$. Our goal is to show that $a_1 = n - 2\gamma + 1$, while $a_2 = \cdots = a_\gamma = 1$, and for this purpose we will use edge rotations (see Lemma 1.1).

Let x^* be the principal eigenvector of G^*. Let $S = \{s_1, \ldots, s_\gamma\}$ be the independent set of G^* such that, for $1 \leq i \leq \gamma$, the vertex s_i has a_i neighbors in the clique K of G^*. Suppose that there exist vertices $s_i, s_j \in S$ such that $a_i, a_j \geq 2$, and without loss of generality, suppose that $x_{s_i}^* \leq x_{s_j}^*$. Let y be an arbitrary vertex adjacent to s_i. By rotating the edge ys_i to ys_j, we get that

$$\lambda_1(G^* - ys_i + ys_j) > \lambda_1(G^*).$$

However, this is a contradiction, as the connected graph

$$G^* - ys_i + ys_j \cong SSG(n, \gamma; a_1, \ldots, a_i - 1, \ldots, a_j + 1, \ldots, a_\gamma)$$

also has domination number γ.

Thus, at most one number among $a_1, \ldots, a_\gamma$ may be larger than 1. This finally shows that

$$G^* \cong S^* \cong SSG(n, \gamma; n - 2\gamma + 1, 1, \ldots, 1).$$

$\square$

When restricted to trees with given domination number, Chen and He [29] proved that

Theorem 4.27 ([29]). *If T is a tree with $n \geq 12$ vertices and domination number γ, then*

$$\lambda_1(T) \leq \lambda_1(T^*),$$

where T^* is a tree obtained from the star $K_{1,n-\gamma}$ by selecting $\gamma - 1$ of its leaves and attaching a new pendant vertex to each of these leaves.

On the opposite side, we can easily characterize the graph G with the minimum spectral radius among graphs with n vertices and the domination number γ, provided we allow disconnected graphs as well. Due to the monotonicity of the spectral radius, such graph G is necessarily a union of γ stars, whose γ center vertices dominate the remaining vertices of G. Since $\lambda_1(G)$ is then equal to the spectral radius of the largest among these γ stars, $\lambda_1(G)$ has minimum possible value if the largest star is the order $\lceil n/\gamma \rceil$. Hence,

$$\lambda_1 \geq \sqrt{\left\lceil \frac{n}{\gamma} \right\rceil - 1}.$$

However, it is still an open problem if we ask which connected graph, or even which tree with n vertices and domination number γ has the minimum spectral radius? It is trivial in the case $\gamma = \lceil n/3 \rceil$, since this is the domination number of the path P_n which has the minimum spectral radis among all n-vertex connected graphs. It has also been solved in the case of trees of even order n and $\gamma = n/2$ by Aouchiche et al. [7], who showed

Theorem 4.28 ([7]). *For even n, the minimum spectral radius among trees with n vertices and the domination number $\gamma = n/2$ is attained by the rooted product of the path $P_{n/2}$ with n copies of K_2.*

Proof of this result relies on the following lemma proved both by Aouchiche et al. [7] and Feng et al. [60]. Recall that $H(G, n)$ denotes the rooted product of graph H with n copies of G (see Section 2.1).

Lemma 4.7. *Let T be a tree with 2k vertices and the domination number k, $k \geq 1$. Then there exists a tree T' such that $T \cong T'(K_2, k)$.*

Proof. By induction on k. If $k = 1$, then $T \cong K_2$ and $T' \cong K_1$.

Next, suppose that the statement holds for all trees with less than $2k$ vertices for some $k \geq 2$, and let T be a tree with $2k$ vertices and the domination number k. We may assume that a dominating set S of cardinality k in T does not contain pendant vertices: if any pendant vertex in S is replaced by its neighbor, the resulting set is still a dominating set.

Let u be a pendant vertex of T, and let v be its only neighbor. By our assumption $v \in S$. We show that u is the unique pendant vertex adjacent to v. Otherwise, let U, $|U| \geq 2$, be the set of all pendant vertices adjacent to v, and let S' be the minimum dominating set of $T - v - U$. Since $T - v - U$ has $2k - 1 - |U|$ vertices, it holds

$$|S'| \leq k - \frac{1 + |U|}{2},$$

but then $S' \cup \{v\}$ is a dominating set of T of cardinality less than k, a contradiction.

Next, the set $S \setminus \{v\}$ is a minimum dominating set in $T - u - v$. In principle, $T - u - v$ consists of connected components $T_1, \ldots, T_l$. Since the domination number γ_i of each T_i is at most half of its order n_i, we conclude that T can have domination number k only if n_i is even $\gamma_i = n_i/2$ for each $i = 1, \ldots, l$. By the inductive hypothesis, for each T_i there exists a tree T'_i such that $T_i = T'_i(K_2, n_i/2)$. The requested tree T' is then obtained by adding v to the union $\cup_{i=1}^{l} T'_i$, and joining v to its non-pendant neighbors in T, which are already contained in $\cup_{i=1}^{l} T'_i$. □

In order to finish the proof of Theorem 4.28, we need to recall from Example 2.1 that if $T = T'(K_2, k)$, then

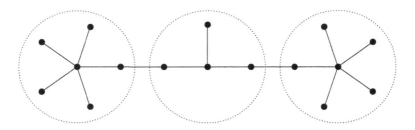

Figure 4.7 The graph with the minimum spectral radius among connected graphs with 16 vertices and domination number 3.

$$\lambda_1(T) = \frac{\lambda_1(T') + \sqrt{\lambda_1^2(T') + 4}}{2}.$$

Hence, T has the minimum spectral radius among trees with $n = 2k$ vertices and domination number k if and only if T' has the minimum spectral radius among trees with k vertices, which happens exactly if $T' = P_k$.

For other values of γ, calculations performed by Alexander Vasilyev with MathChem [158] on the author's request, suggest the following

Conjecture 4.13. *If $\gamma \leq \lceil n/3 \rceil$, then the minimum spectral radius among connected graphs with n vertices and domination number γ is attained by the tree obtained from a set of γ stars joined in a path-like structure (see Fig. 4.7), such that the orders of stars form an inversely unimodal sequence, decreasing from both end stars toward the middle star.*

4.9 NORDHAUS-GADDUM INEQUALITY FOR THE SPECTRAL RADIUS

A Nordhaus-Gaddum inequality is an inequality that provides a bound on the value of $i(G) + i(\overline{G})$ for some graph invariant i, with the bound usually given as a function of the number n of vertices of G. There are quite a few such inequalities for various graph invariants, which are extensively surveyed in [6]. The first such inequalities were proved for the chromatic number by Nordhaus and Gaddum [115] in 1956.

Theorem 4.29 ([115]). *For a graph G with n vertices*

$$2\sqrt{n} \leq \chi(G) + \chi(\overline{G}) \leq n + 1,$$

$$n \leq \chi(G) \cdot \chi(\overline{G}) \leq \frac{(n+1)^2}{4}.$$

Nosal has obtained such inequalities for the spectral radius in her Master's thesis [116] in 1970.

Theorem 4.30 ([116]). $n - 1 \leq \lambda_1(G) + \lambda_1(\overline{G}) \leq n\sqrt{2}$.

The lower bound is attained if and only if G is regular, while the absence of graphs attaining the upper bound above has motivated the search for a sharp upper bound. Hong and Shu [83] obtained an upper bound in terms of the number n of vertices and the chromatic number χ, and Nikiforov [110] improved it by replacing the chromatic number with the clique number ω.

Theorem 4.31 ([83]). *For a graph G with n vertices*

$$\lambda_1(G) + \lambda_1(\overline{G}) \leq \sqrt{\left(2 - \frac{1}{\chi(G)} - \frac{1}{\chi(\overline{G})}\right) n(n-1)},$$

with equality if and only if G is the complete graph K_n or the empty graph $\overline{K_n}$.

Theorem 4.32 ([110]). *For a graph G with n vertices*

$$\lambda_1(G) + \lambda_1(\overline{G}) \leq \sqrt{\left(2 - \frac{1}{\omega(G)} - \frac{1}{\omega(\overline{G})}\right) n(n-1)},$$

with equality if and only if G is the complete graph K_n or the empty graph $\overline{K_n}$.

Proof. From Theorem 4.1 we have that

$$\lambda_1(G) + \lambda_1(\overline{G}) \leq \sqrt{2m\left(1 - \frac{1}{\omega(G)}\right)} + \sqrt{[n(n-1) - 2m]\left(1 - \frac{1}{\omega(\overline{G})}\right)}.$$

For fixed n, $\omega(G)$ and $\omega(\overline{G})$, the right-hand side above may be treated as a function in m:

$$f(m) = \sqrt{2m\left(1 - \frac{1}{\omega(G)}\right)} + \sqrt{[n(n-1) - 2m]\left(1 - \frac{1}{\omega(\overline{G})}\right)}.$$

Its derivative is

$$f'(m) = \frac{\sqrt{[n(n-1) - 2m]\left(1 - \frac{1}{\omega(G)}\right)} - \sqrt{2m\left(1 - \frac{1}{\omega(\overline{G})}\right)}}{\sqrt{m[n(n-1) - 2m]}},$$

and it is easy to see that its only stationary point is

$$m^* = \frac{1 - \frac{1}{\omega(G)}}{2 - \frac{1}{\omega(G)} - \frac{1}{\omega(\overline{G})}}\binom{n}{2},$$

which, together with $f'(0) > 0$ and $f'\left(\binom{n}{2}\right) < 0$, shows that f has the maximum value at m^* in the interval $\left[0, \binom{n}{2}\right]$. Hence

$$\lambda_1(G) + \lambda_1(\overline{G}) \leq f(m) \leq f(m^*) = \sqrt{\left(2 - \frac{1}{\omega(G)} - \frac{1}{\omega(\overline{G})}\right)n(n-1)}.$$

The case of equality follows from the Motzkin-Straus Lemma 4.1, which implies that the principal eigenvector components have to be mutually equal at the vertices belonging to the maximum clique, and zero elsewhere. As either G or $\overline{G}$ is connected, and hence has a strictly positive principal eigenvector, we conclude that either G or $\overline{G}$ has to be the complete graph K_n. $\square$

Unfortunately, bounds from the two previous theorems are not much better than Nosal's upper bound when both $\chi(G)$ and $\chi(\overline{G})$ or $\omega(G)$ and $\omega(\overline{G})$ are large enough. Experiments with AutoGraphiX have led Aouchiche et al. [3] to Conjecture 4.3, while Nikiforov [109] has independently proposed a similar conjecture with the same coefficient of n, although without conjecturing the extremal graph.

Conjecture 4.14 ([109]). $\lambda_1(G) + \lambda_1(\overline{G}) \leq \frac{4}{3}n + O(1)$.

This conjecture has been attacked with some success by Csikvári [40], who proved

Theorem 4.33 ([40]). $\lambda_1(G) + \lambda_1(\overline{G}) \leq \frac{1 + \sqrt{3}}{2}n - 1$.

There are two main ingredients of Csikvári's proof: a simple upper bound on the spectral radius of graphs reminiscent of the complete split graphs, and the Kelmans transformation that turns out to simultaneously increase the spectral radii of both G and $\overline{G}$.

Lemma 4.8 ([40]). *Assume that the set $K = \{v_1, \ldots, v_k\}$ forms a clique in the graph G, while the set $V(G) \backslash K = \{v_{k+1}, \ldots, v_n\}$ forms an independent set. If e denotes the number of edges between a vertex in K and a vertex in $V(G) \backslash K$, then*

$$\lambda_1 \leq \frac{k - 1 + \sqrt{(k-1)^2 + 4e}}{2}.$$

Proof. Let x be the principal eigenvector of G. For $1 \leq j \leq k$ we have

$$\lambda_1 x_{v_j} = \sum_{\substack{1 \leq i \leq k \\ i \neq j}} x_{v_i} + \sum_{v_l \in N_{v_j} \cap (V(G) \backslash K)} x_{v_l}.$$

By adding together these eigenvalue equations we get

$$\lambda_1 \left(\sum_{i=1}^{k} x_{v_i} \right) = (k-1) \left(\sum_{i=1}^{k} x_{v_i} \right) + \sum_{l=k+1}^{n} \deg_{v_l} x_{v_l}. \tag{4.32}$$

Since $V(G) \backslash K$ forms an independent set, all the neighbors of v_l for $k + 1 \leq l \leq n$ are among $v_1, \ldots, v_k$, so that $\lambda_1 x_{v_l} \leq \sum_{i=1}^{k} x_{v_i}$. From (4.32) we now get

$$\lambda_1 \left(\sum_{i=1}^{k} x_{v_i} \right) \leq (k-1) \left(\sum_{i=1}^{k} x_{v_i} \right) + \sum_{l=k+1}^{n} \frac{\deg_{v_l}}{\lambda_1} \left(\sum_{i=1}^{k} x_{v_i} \right).$$

Dividing by $\sum_{i=1}^{k} x_{v_i}$ and noting that $\sum_{l=k+1}^{n} \deg_{v_l} = e$, we get

$$\lambda_1 \leq k - 1 + \frac{e}{\lambda_1},$$

and by solving this quadratic inequality it follows that

$$\lambda_1 \leq \frac{k - 1 + \sqrt{(k-1)^2 + 4e}}{2}.$$

$\square$

The second ingredient of Csikvári's proof is the transformation, introduced by Kelmans in [85]. If u and v are two vertices of G, the Kelmans transform of G is obtained by removing all edges between u and

$N_u \setminus (\{v\} \cup N_v)$, and adding all edges between v and $N_u \setminus (\{v\} \cup N_v)$. We will call v the beneficiary of the transformation.

Let G' be the Kelmans transform of G. Observe firstly that G' is, up to isomorphism, independent of the choice of the beneficiary among u and v: in G' one of u and v will necessarily be adjacent to vertices in $N_u \cup N_v$, while the other will be adjacent to vertices in $N_u \cap N_v$ (and the subgraph $G - u - v$ does not change).

Let λ_1 and x be the spectral radius and the principal eigenvector of G, and let A_G and $A_{G'}$ be the adjacency matrices of G and G', respectively. Assume that $x_u \geq x_v$ and choose u to be the beneficiary of the Kelmans transformation (which does not affect the resulting graph G'). Then by the Rayleigh quotient

$$\lambda_1(G') \geq x^T A_{G'} x = x^T A_G x + 2(x_u - x_v) \sum_{w \in N_v \setminus (\{u\} \cup N_u)} x_w \geq \lambda_1(G).$$

Since the Kelmans transformation of G into G' is, at the same time, the Kelmans transformation of $\overline{G}$ into $\overline{G}'$ with the other vertex becoming the beneficiary, it immediately follows that also

$$\lambda_1(\overline{G}') \geq \lambda_1(\overline{G}).$$

Hence, the Kelmans transformation increases spectral radii of both G and $\overline{G}$.

Further, let us say that u dominates v if all the neighbors of v, except possibly u, are also the neighbors of u, i.e., if $N_v \setminus \{u\} \subset N_u \setminus \{v\}$. Clearly if u is the beneficiary of the Kelmans transformation, then u will dominate v in G'.

Let $\deg_1^G \geq \deg_2^G \geq \cdots \geq \deg_n^G$ be the nonincreasing degree sequence of G. Introduce partial ordering $\prec$ such that $G_1 \prec G_2$ if for some k holds $\deg_k^{G_1} < \deg_k^{G_2}$, while $\deg_i^{G_1} = \deg_i^{G_2}$ for $1 \leq i \leq k - 1$. Choose G^* to be a maximal graph for $\prec$ among all graphs that can be obtained from G by a sequence of Kelmans transformations. We claim that when the vertices of G^* are ordered such that $\deg_{v_1}^{G^*} \geq \cdots \geq \deg_{v_n}^{G^*}$, then v_i dominates v_j whenever $i < j$. Namely, if $\deg_{v_i}^{G^*} \geq \deg_{v_j}^{G^*}$, but v_i does not dominate v_j, then one can apply the Kelmans transformation to G^*, v_i, and v_j, with v_i as its beneficiary; the degree of v_i in the resulting graph G^{**} will be strictly larger than $\deg_{v_i}^{G^*}$, so that $G^* \prec G^{**}$, a contradiction.

Now, in order to finish the proof of Theorem 4.33, it is enough to show that

$$\lambda_1(G^*) + \lambda_1(\overline{G^*}) \le \frac{1 + \sqrt{3}}{2}n.$$

Let k be the smallest integer for which v_k and v_{k+1} are not adjacent. Then $K = \{v_1, \ldots, v_k\}$ forms a clique, while $V(G^*) \setminus K = \{v_{k+1}, \ldots, v_n\}$ forms an independent set in G^*. Let e be the number of edges between a vertex in K and a vertex in $V(G^*) \setminus K$. From Lemma 4.8 we have

$$\lambda_1(G^*) \le \frac{k - 1 + \sqrt{(k-1)^2 + 4e}}{2},$$

$$\lambda_1(\overline{G^*}) \le \frac{n - k - 1 + \sqrt{(n-k-1)^2 + 4[k(n-k) - e]}}{2},$$

so that

$$2\lambda_1(G^*) + 2\lambda_1(\overline{G^*}) - (n - 2)$$
$$\le \sqrt{(k-1)^2 + 4e} + \sqrt{(n-k-1)^2 + 4[k(n-k) - e]}. \quad (4.33)$$

From the inequality between arithmetic and square mean of two numbers, we get

$$\sqrt{(k-1)^2 + 4e} + \sqrt{(n-k-1)^2 + 4[k(n-k) - e]}$$

$$\le \sqrt{2\left[(k-1)^2 + (n-k-1)^2 + 4k(n-k)\right]}$$

$$\le \sqrt{2\left[2\left(\frac{n}{2} - 1\right)^2 + 4\frac{n^2}{4}\right]} < n\sqrt{3}.$$

Hence

$$2\lambda_1(G^*) + 2\lambda_1(\overline{G^*}) - (n - 2) < n\sqrt{3},$$

so that

$$\lambda_1(G^*) + \lambda_1(\overline{G^*}) < \frac{1 + \sqrt{3}}{2}n - 1.$$

The term $\frac{1+\sqrt{3}}{2}n \approx 1.3660\,n$ in the inequality above was successfully lowered by Terpai [149] to the value of $\frac{4}{3}n$, as conjectured by Nikiforov [109]. In his proof, Terpai has resorted to graphons—the limit objects of convergent sequences of dense graphs—which are beyond the scope of this book. The reader is referred to Lovász's book [97] for introduction to graphons, and to Terpai's paper [149] itself.

Note, however, that Terpai's result does not attempt to characterize extremal graphs for $\lambda_1(G) + \lambda_1(\overline{G})$. Recall that Conjecture 4.3 proposes that such graphs are the complete split graph $CS_{n,\lceil 2n/3 \rceil}$ and its complement, and if $n \equiv 2 \pmod 3$, also the complete split graph $CS_{n,\lfloor 2n/3 \rfloor}$ and its complement. Interestingly, Csikvári's Lemma 4.8 is sharp exactly for complete split graphs and their complements, i.e., for $e = 0$ and $e = k(n-k)$. However, the value (4.33), when considered as a function of e for fixed values of n and k, is a concave function on the interval $[0, k(n - k)]$, so that it reaches the maximum value at an interior point of this interval, where Lemma 4.8 is no longer sharp. It remains to be seen whether Conjecture 4.3 could be proved by either improving Lemma 4.8 or by exploiting further properties of the $\prec$-maximal graph G^*.

BIBLIOGRAPHY

[1] Alon N. A note on degenerate and spectrally degenerate graphs. J Graph Theory 2013;72:1–6.

[2] Andjelić M, da Fonseca CM, Simić SK, Tošić DV. On bounds for the index of double nested graphs. Linear Algebra Appl 2011;435:2475–90.

[3] Aouchiche M, Bell FK, Cvetković D, Hansen P, Rowlinson P, Simić S, Stevanović D. Variable neighborhood search for extremal graphs. 16: Some conjectures related to the largest eigenvalue of a graph. Eur J Oper Res 2008;191:661–76.

[4] Aouchiche M, Caporossi G, Hansen P. Open problems on graph eigenvalues studied with AutoGraphiX. Eur J Comput Optim 2013;1:181–99.

[5] Aouchiche M, Hansen P. A survey of automated conjectures in spectral graph theory. Linear Algebra Appl 2010;432:2293–322.

[6] Aouchiche M, Hansen P. A survey of Nordhaus-Gaddum type relations. Discrete Appl Math 2013;161:466–546.

[7] Aouchiche M, Hansen P, Stevanović D. Variable neighborhood search for extremal graphs. 17: Further conjectures and results about the index. Discuss Math Graph Theory 2009;29:15–37.

[8] Belardo F, Li Marzi EM, Simić S. Trees with minimal index and diameter at most four. Discrete Math 2010;310:1708–14.

[9] Belardo F, Li Marzi EM, Simić S. Bidegreed trees with small index. MATCH Commun Math Comput Chem 2009;61:503–15.

[10] Belardo F, Li Marzi EM, Simić S, Wang J. On the spectral radius of unicyclic graphs with prescribed degree sequence. Linear Algebra Appl 2010;432:2323–34.

[11] Bell FK. On the maximal index of connected graphs. Linear Algebra Appl 1991;144:135–51.

[12] Bell FK. A note on the irregularity of graphs. Linear Algebra Appl 1992;161:45–54.

[13] Berge C. Sur le couplage maximum d'un graphe. CR Acad Sci Paris 1958;247:258–9 [in French].

[14] Bermond JC, Fournier JC, Las Vergnas M, Sotteau D, editors. Problémes combinatoires et theories des graphes. Coll Int CNRS No. 260, 1976. Orsay: CNRS; 1978.

[15] Bhattacharya A, Friedland S, Peled UN. On the first eigenvalue of bipartite graphs. Electron J Comb 2008;15:144.

[16] Biggs NL, Mohar B, Shawe-Taylor J. The spectral radius of infinite graphs. Bull Lond Math Soc 1988;20:116–20.

[17] Bıyıkoğlu T, Leydold J. Graphs with given degree sequence and maximal spectral radius. Electron J Comb 2008;15:#R119.

[18] Bıyıkoğlu T, Leydold J. Dendrimers are the unique chemical trees with maximum spectral radius. MATCH Commun Math Comput Chem 2012;68:851–4.

[19] Bollobás B. Extremal graph theory. New York: Courier Dover Publishers; 2004.

[20] Boots BN, Royle GF. A conjecture on the maximum value of the principal eigenvalue of a planar graph. Geographical Anal 1991;23:276–82.

[21] Brouwer AE, Haemers WH. Spectra of graphs. New York: Springer; 2012.

147

[22] Brualdi RA. Spectra of digraphs. Linear Algebra Appl 2010;432:2181–13.

[23] Brualdi RA, Hoffman AJ. On the spectral radius of 0,1 matrices. Linear Algebra Appl 1985;65:133–46.

[24] Brualdi RA, Solheid ES. On the spectral radius of complementary acyclic matrices of zeros and ones. SIAM J Algebra Discrete Method 1986;7:265–72.

[25] Brualdi RA, Solheid ES. On the spectral radius of connected graphs. Publ Inst Math Belgrade NS 1986;39(53):45–54.

[26] Cao D, Vince A. The spectral radius of a planar graph. Linear Algebra Appl 1993;187:251–7.

[27] Caporossi G, Hansen P. Variable neighborhood search for extremal graphs. I. The AutoGraphiX system. Discrete Math 2000;212:29–44.

[28] Caporossi G, Hansen P. Variable neighborhood search for extremal graphs. V. Three ways to automate finding conjectures. Discrete Math 2004;276:81–94.

[29] Chen P, He C. On the spectral radii of trees with given domination number. Adv Math Beijing 2012;41:225–32.

[30] Chrobak M, Eppstein D. Planar orientations with low out-degree and compaction of adjacency matrices. Theor Comput Sci 1991;86:243–66.

[31] Cioabă SM. The spectral radius and the maximum degree of irregular graphs. Electron J Comb 2007;14:#R38.

[32] Cioabă SM. A necessary and sufficient eigenvector condition for a connected graph to be bipartite. Electron J Linear Algebra 2010;20:351–3.

[33] Cioabă SM, Gregory DA. Principal eigenvectors of irregular graphs. Electron J Linear Algebra 2007;16:366–79.

[34] Cioabă SM, Gregory DA. Large matchings from eigenvalues. Linear Algebra Appl 2007;422:308–17.

[35] Cioabă SM, Gregory DA, Nikiforov V. Extreme eigenvalues of nonregular graphs. J Comb Theory Ser B 2007;97:483–6.

[36] Cioabă SM, van Dam ER, Koolen JH, Lee JH. A lower bound for the spectral radius of graphs with fixed diameter. Eur J Comb 2010;31:1560–6.

[37] Cioabă SM, van Dam ER, Koolen JH, Lee JH. Asymptotic results on the spectral radius and the diameter of graphs. Linear Algebra Appl 2010;432:722–37.

[38] Collatz L, Sinogowitz U. Spektren endlicher graphen. Abh Math Sem Univ Hamburg 1957;21:63–77.

[39] Cormen TH, Leiserson CE, Rivest RL, Stein C. Introduction to algorithms. Boston: MIT Press; 2001.

[40] Csikvári P. On a conjecture of V. Nikiforov. Discrete Math 2009;309:4522–6.

[41] Cull P, Flahive M, Robson R. Difference equations: from rabbits to chaos. New York: Springer; 2005.

[42] Cvetković D. Chromatic number and the spectrum of a graph. Publ Inst Math Belgrade NS 1972;14(28):25–38.

[43] Cvetković D, Doob M, Sachs H. Spectra of graphs–theory and application. New York: Academic Press; 1980.

[44] Cvetković D, Hansen P, Kovačević-Vujčić V. On some interconnections between combinatorial optimization and extremal graph theory. Yugoslav J Oper Res 2004;14:147–54.

[45] Cvetković D, Rowlinson P. On connected graphs with maximal index. Publ Inst Math Belgrade NS 1988;44(58):29–34.

[46] Cvetković D, Rowlinson P, Simić S. Eigenspaces of graphs. Cambridge: Cambridge University Press; 1997.

[47] Cvetković D, Rowlinson P, Simić S. An introduction to the theory of graph spectra. New York: Cambridge University Press; 2009.

[48] Cvetković D, Simić S. Graph theoretical results obtained by the support of the expert system "Graph" –an extended survey. In: Fajtlowicz S, Fowler PW, Hansen P, Janowitz MF, Roberts FS, editors. Graphs and discovery. Providence: American Mathematical Society; 2005, p. 39–70.

[49] Delorme C. Eigenvalues of complete multipartite graphs. Discrete Math 2012;312:2532–5.

[50] Diestel R. Graph theory. 4th ed. Heidelberg: Springer-Verlag; 2010.

[51] Dress A, Gutman I. The number of walks in a graph. Appl Math Lett 2003;16:797–801.

[52] Dress A, Stevanović D. Hoffman-type identities. Appl Math Lett 2003;16:297–302.

[53] Dress A, Stevanović D. A note on a theorem of Horst Sachs. Ann Comb 2004;8:487–97.

[54] Dvořák Z, Mohar B. Spectral radius of finite and infinite planar graphs and of graphs of bounded genus. J Comb Theory Ser B 2010;100:729–39.

[55] Dvořák Z, Mohar B. Spectrally degenerate graphs: hereditary case. J Comb Theory Ser B 2012;102:1099–109.

[56] Du X, Shi L. Graphs with small independence number minimizing the spectral radius. Discrete Math Algorithms Appl 2013;5:1350017.

[57] Edwards CS, Elphick CH. Lower bounds for the clique and the chromatic numbers of a graph. Discrete Appl Math 1983;5:51–64.

[58] Fajtlowicz S. On conjectures of graffiti. Discrete Math 1988;72:113–18.

[59] Feng L, Li Q, Zhang XD. Spectral radii of graphs with given chromatic number. Appl Math Lett 2007;20:158–62.

[60] Feng L, Li Q, Zhang XD. Minimizing the Laplacian eigenvalues for trees with given domination number. Linear Algebra Appl 2006;419:648–55.

[61] Feng L, Yu G, Zhang XD. Spectral radius of graphs with given matching number. Linear Algebra Appl 2007;422:133–8.

[62] Fisher RA. An examination of the different possible solutions of a problem in incomplete blocks. Ann Eugenics 1940;10:52–75.

[63] Friedland S. Matrices. http://homepages.math.uic.edu/~friedlan/bookm.pdf. Cited 26 June 2014.

[64] Friedman J. The spectra of infinite hypertrees. SIAM J Comput 1991;20:951–61.

[65] Gantmacher FR. The theory of matrices. vol. II. New York: Chelsea Publishing Company; 1959.

[66] Gantmacher FR, Krein MG. Oscillation matrices and kernels and small vibrations of mechanical systems. Translation Series AEC-tr-4481. Washington, DC: USAEC Office of Technical Information; 1961.

[67] Garey MR, Johnson DS. Computers and intractability: a guide to the theory of NP-completeness. New York: W.H. Freeman; 1979.

[68] Godsil CD. Spectra of trees. Ann Discrete Math 1984;20:151–9.

[69] Godsil CD, McKay BD. A new graph product and its spectrum. Bull Austral Math Soc 1978;18:21–28.

[70] Gonçalves D. Covering planar graphs with forests, one having bounded maximum degree. J Comb Theory Ser B 2009;99:314–22.

[71] Gregory AD, Hershkovitz D, Kirkland SJ. The spread of the spectrum of a graph. Linear Algebra Appl 2001;332–334:23–35.

[72] Hansen P, Stevanović D. On bags and bugs. Discrete Appl Math 2008;156:986–97.

[73] Harary F, King C, Mowshowitz A, Read RC. Cospectral graphs and digraphs. Bull. Lond Math Soc 1971;3:321–8.

[74] Harary F, Thomassen C. Anticritical graphs. Math Proc Camb Phil Soc 1976;79:11–18.

[75] Hayes TP. A simple condition implying rapid mixing of single-site dynamics on spin systems. In: Proc. 47th Annual IEEE Symposium on Foundations of Computer Science FOCS 2006. New York: IEEE; 2006, p. 39–46.

[76] Heilbronner E. Das Kompositions-Prinzip: Eine anschauliche Methode zur elektronentheoretischen Behandlung nicht oder niedriger symmetrischer Molekeln im Rahmen der MO-Theorie. Helv Chim Acta 1953;36:170–88.

[77] Heilmann OJ, Lieb EH. Theory of monomer-dimer systems. Commun Math Phys 1972;25: 190–232.

[78] Hoffman AJ. On the polynomial of a graph. Amer Math Monthly 1963;70:30–36.

[79] Hoffman AJ, Singleton RR. Moore graphs with diameter 2 and 3. IBM J Res Dev 1960;4:497–504.

[80] Hoffman AJ, Smith JH. On the spectral radii of topologically equivalent graphs. In: Fiedler M, editor. Recent advances in graph theory. Prague: Academia Praha; 1975, p. 273–81.

[81] Hofmeister M. Spectral radius and degree sequence. Math Nachr 1988;139:37–44.

[82] Hong Y. A bound on the spectral radius of graphs. Linear Algebra Appl 1988;108:135–9.

[83] Hong Y, Shu J. A sharp upper bound for the spectral radius of the Nordhaus-Gaddum type. Discrete Math 2000;211:229–32.

[84] Hu S. The largest eigenvalue of unicyclic graphs. Discrete Math 2007;307:280–4.

[85] Kelmans AK. On graphs with randomly deleted edges. Acta Math Acad Sci Hung 1981; 37:77–88.

[86] Krivelevich M. An improved upper bound on the minimal number of edges in color-critical graphs. Electron J Comb 1998;1:R4.

[87] Lan J, Li L, Shi L. Graphs with diameter $n - e$ minimizing the spectral radius. Linear Algebra Appl 2012;437:2823–50.

[88] Lan J, Lu L. Diameters of graphs with spectral radius at most $\frac{3}{2}\sqrt{2}$. Linear Algebra Appl 2013;438:4382–407.

[89] Li Q, Feng K. On the largest eigenvalue of a graph. Acta Math Appl Sinica 1979;2:167–75 [in Chinese].

[90] Li C, Wang H, Van Mieghem P. Bounds for the spectral radius of a graph when nodes are removed. Linear Algebra Appl 2012;437:319–23.

[91] Liu B, Li G. A note on the largest eigenvalue of non-regular graphs. Electron J Linear Algebra 2008;17:54–61.

[92] Liu M, Liu B. Some results on the majorization theorem of connected graphs. Acta Math Sin Engl Ser 2012;28:371–8.

[93] Liu M, Liu B, You Z. The majorization theorem of connected graphs. Linear Algebra Appl 2009;431:553–7.

[94] Liu H, Lu M, Tian F. On the spectral radius of unicyclic graphs with fixed diameter. Linear Algebra Appl 2007;420:449–57.

[95] Liu B, Mu-Huo L, You Z. Erratum to 'A note on the largest eigenvalue of non-regular graphs.' Electron J Linear Algebra 2009;18:64–68.

[96] Liu B, Shen J, Wang X. On the largest eigenvalue of non-regular graphs. J Comb Theory Ser B 2007;97:1010–18.

[97] Lovász L. Large networks and graph limits. Providence: American Mathematical Society; 2012.

[98] Lovász L, Pelikán J. On the eigenvalues of trees. Period Math Hungar 1973;3:175–82.

[99] Mahadev NVR, Peled UN. Threshold graphs and related topics. New York: North-Holland; 1995.

[100] Majstorović S, Stevanović D. Graphs with the largest eigenvalue of modularity matrix equal to zero. Electron J Linear Algebra, in press.

[101] Meyer CD. Matrix analysis and applied linear algebra. Philadelphia: Society for Industrial and Applied Mathematics; 2000.

[102] Miller M, Širáň J. Moore graphs and beyond: a survey of the degree/diameter problem. Electron J Comb 2013;20:#DS14v2.

[103] Mladenović N, Hansen P. Variable neighborhood search. Comput Oper Res 1997;24:1097–100.

[104] Mohar B. The Spectrum of an Infinite Graph. Linear Algebra Appl 1982;48:245–56.

[105] Mohar B, Omladič M. Divisors and the spectrum of infinite graphs. Linear Algebra Appl 1987;91:99–106.

[106] Mohar B, Woess W. A survey on spectra of infinite graphs. Bull Lond Math Soc 1989;21: 209–34.

[107] Motzkin TS, Straus EG. Maxima for graphs and a new proof of a theorem of Turán. Canad J Math 1965;17:533–40.

[108] Nemwan MEJ. Networks: an introduction. New York: Oxford University Press; 2010.

[109] Nikiforov V. Eigenvalue problems of Nordhaus-Gaddum type. Discrete Math 2007;307:774–80.

[110] Nikiforov V. Some inequalities for the largest eigenvalue of a graph. Comb Prob Comput 2002;11:179–89.

[111] Nikiforov V. Bounds on graph eigenvalues. II. Linear Algebra Appl 2007;427:183–9.

[112] Nikiforov V. The spectral radius of subgraphs of regular graphs. Electron J Comb 2007;14:#N20.

[113] Nikiforov V. A spectral condition for odd cycles in graphs. Linear Algebra Appl 2008;428: 1492–8.

[114] Nikiforov V. Some new results in extremal graph theory. In: Chapman R, editor. Surveys in combinatorics 2011. Cambridge: Cambridge University Press; 2011, p. 141–82.

[115] Nordhaus EA, Gaddum J. On complementary graphs. Amer Math Monthly 1956;63:175–7.

[116] Nosal E. Eigenvalues of graphs. Masters Thesis. University of Calgary, 1970.

[117] Olesky DD, Roy A, van den Driessche P. Maximal graphs and graphs with maximal spectral radius. Linear Algebra Appl 2002;346:109–30.

[118] Ore O. Theory of graphs. Providence: American Mathematical Society; 1962.

[119] Ostrowski A. On the eigenvector belonging to the maximal root of a nonnegative matrix. Proc Edinburgh Math Soc 1960/1961;12:107–12.

[120] Papendieck B, Recht P. On maximal entries in the principal eigenvector of graphs. Linear Algebra Appl 2000;310:129–38.

[121] Peters A, Coolsaet K, Brinkmann G, Van Cleemput N, Fack V. GrInvIn in a nutshell. J. Math. Chem. 2009;45:471–7.

[122] Petrović MM. The spectrum of infinite complete multipartite graphs. Publ Inst Math Belgrade NS 1982;31(45):169–76.

[123] Pruitt WE. Eigenvalues of non-negative matrices. Ann Math Stat 1964;35:1797–800.

[124] Pruss AR. Discrete convolution-rearrangement inequalities and the Faber-Krahn inequality on regular trees. Duke Math J 1998;91:463–514.

[125] Rojo O, Robbiano M. An explicit formula for eigenvalues of Bethe trees and upper bounds on the largest eigenvalue of any tree. Linear Algebra Appl 2007;427:138–50.

[126] Rowlinson P. On the maximal index of graphs with a prescribed number of edges. Linear Algebra Appl. 1988;110:43–53.

[127] Rowlinson P. More on graph perturbations. Bull Lond Math Soc 1990;22:209–16.

[128] Sachs H. Beziehungen zwischen den in einem Graphen enthaltenen Kreisen und seinem charakteristischen Polynom. Publ Math Debrecen 1964;11:119–34.

[129] Sanchis L. Maximum number of edges in connected graphs with a given domination number. Discrete Math 1991;87:65–72.

[130] Schwenk AJ. Almost all trees are cospectral. In: Harary, F. editor. New directions in the theory of graphs. New York: Academic Press; 1973, p. 275–307.

[131] Schwenk AJ. Computing the characteristic polynomial of a graph. In: Proc. capital conf. graph theory and combinatorics. Lecture Notes in Mathematics 406. Berlin: Springer; 1974, p. 153–72.

[132] Schwenk AJ, Wilson RJ. On the eigenvalues of a graph. In: Beineke LW, Wilson RJ, editors. Selected topics in graph theory. New York: Academic Press; 1978, p. 307–36.

[133] Seneta E. Non-negative matrices and Markov chains. Berlin: Springer; 1981.

[134] Shi L. The spectral radius of irregular graphs. Linear Algebra Appl 2009;431:189–96.

[135] Simić S, Li Marzi EM, Belardo F. On the index of caterpillars. Discrete Math 2008;308:324–30.

[136] Simić SK, Tošić DV. The index of trees with specified maximum degree. MATCH Commun. Math. Comput. Chem. 2005;54:351–62.

[137] Smith JH. Some properties of the spectrum of a graph. In: Guy R, Hanani H, Sauer N, Schönheim J, editors. Combinatorial structures and their applications. New York: Gordon & Breach; 1970, p. 403–406.

[138] Song H, Wang Q, Tian L. New upper bounds on the spectral radius of trees with the given number of vertices and maximum degree. Linear Algebra Appl 2013;439:2527–41.

[139] Stevanović D. Resolution of AutoGraphiX conjectures relating the index and matching number of graphs. Linear Algebra Appl 2010;433:1674–7.

[140] Stevanović D. Research problems from the Aveiro workshop on graph spectra. Linear Algebra Appl 2007;423:172–81.

[141] Stevanović D. The largest eigenvalue of nonregular graphs. J Comb Theory Ser B 2004;91: 143–6.

[142] Stevanović D. Bounding the largest eigenvalue of trees in terms of the largest vertex degree. Linear Algebra Appl 2003;360:35–42.

[143] Stevanović D, Aouchiche M, Hansen P. On the spectral radius of graphs with a given domination number. Linear Algebra Appl 2008;428:1854–64.

[144] Stevanović D, Brankov V. An invitation to newGRAPH, Rend Semin Mat Messina Ser II 2004;9(25):211–16.

[145] Stevanović D, Gutman I, Rehman MU. On spectral radius and energy of complete multipartite graphs. Ars Math Contemp 2015;9:109–13.

[146] Stevanović D, Hansen P. The minimum spectral radius of graphs with a given clique number. Electron J Linear Algebra 2008;17:110–17.

[147] Stevanović D, Ilić A. Distance spectral radius of trees with fixed maximum degree. Electron J Linear Algebra 2010;20:168–79.

[148] Sumner DP, Blitch P. Domination critical graphs. J Comb Theory Ser B 1983;34:65–76.

[149] Terpai T. Proof of a conjecture of V. Nikiforov. Combinatorica 2011;31:739–54.

[150] Thompson RC. The behavior of eigenvalues and singular value under perturbations of restricted rank. Linear Algebra Appl 1976;13:69–78.

[151] Turán P. On an extremal problem in graph theory. Mat Fiz Lap 1941;48:436–52 [in Hungarian].

[152] Tutte WT. The factorization of linear graphs. J Lond Math Soc 1947;22:107–11.

[153] van Dam ER. Graphs with given diameter maximizing the spectral radius. Linear Algebra Appl 2007;426:454–7.

[154] van Dam ER, Kooij RE. The minimal spectral radius of graphs with a given diameter. Linear Algebra Appl 2007;423:408–19.

[155] Van Mieghem P. Graph spectra for complex networks. New York: Cambridge University Press; 2011.

[156] Van Mieghem P, Omić J, Kooij RE. Virus spread in networks. IEEE/ACM Trans Netw 2009;17:1–14.

[157] Van Mieghem P, Stevanović D, Kuipers F, Li C, van de Bovenkamp R, Lu D, Wang H. Decreasing the spectral radius of a graph by link removals. Phys Rev E 2011;84:016101.

[158] Vasilyev A, Stevanović D. MathChem: a Python package for calculating topological indices. MATCH Commun Math Comput Chem 2014;71:657–80.

[159] Vizing VG. On the number of edges in a graph with given radius. (In Russian.) Dokl Akad Nauk SSSR 1967;173:1245–6.

[160] Wang J, Huang Q, An X, Belardo F. Some notes on graphs whose spectral radius is close to $\frac{3}{2}\sqrt{2}$. Linear Algebra Appl 2008;429:1606–18.

[161] Weyl H. Das asymptotische verteilungsgesetz der eigenwerte linearer partieller differentialgleichungen. Math Ann 1912;71:441–79.

[162] Wilf HS. The eigenvalues of a graph and its chromatic number. J Lond Math Soc 1967;42:330–2.

[163] Wilf HS. Spectral bounds for the clique and independence numbers of graphs. J Comb Theory Ser B 1986;40:113–17.

[164] Woo R, Neumaier A. On graphs whose spectral radius is bounded by $\frac{3}{2}\sqrt{2}$. Graphs Comb 2007;23:713–26.

[165] Xu M, Hong Y, Shu J, Zhai M. The minimum spectral radius of graphs with a given independence number. Linear Algebra Appl 2009;431:937–45.

[166] Yuan XY, Shao JY, Liu Y. The minimal spectral radius of graphs of order n with diameter $n - 4$. Linear Algebra Appl 2008;428:2840–51.

[167] Zhai W, Liu R, Shu J. On the spectral radius of bipartite graphs with given diameter. Linear Algebra Appl 2009;430:1165–70.

[168] Zhang XD. Eigenvectors and eigenvalues of non-regular graphs. Linear Algebra Appl 2005;409:79–86.

[169] Zykov AA. On some properties of linear complexes. Mat Sbornik NS 1949;24(66):163–88 [in Russian].

INDEX

adjacency matrix, 4
  stepwise, 48
adjacency operator, 25
approximation, 41
AutoGraphiX, 105, 121, 129

balanced incomplete block design, 114
BFD-ordering, 65
breadth-first search, 64

caterpillar, 73
characteristic polynomial, 7, 84, 110
Chebyshev polynomial, 8
chromatic number, 2, 147
clique, 2
clique number, 2, 147
coalescence, 3, 7
coloring, 2
complete
  bipartite graph, 8
  graph, 8
conjecture
  irregularity, 89
  limit of min λ_1 fixed D, 115
  max λ_1 fixed degree sequence, 68
  max λ_1 for planar graphs, 75
  min λ_1 fixed α, 100
  min λ_1 for trees with fixed γ, 146
  min λ_1 when $D \approx 2n/3$, 120
  non-regular graphs, 63
  Nordhaus-Gaddum, 90
  principal ratio, 37
  spectral spread, 90
connected vertices, 1
cut edge, 7
cycle, 3, 8

diameter, 2
distance, 2
dominating set, 2
domination number, 2

eccentricity, 2
edge
  orientation, 76
  rotation, 10
  set, 1
  switching, 10
eigenvalue, 5
  equation, 5

multiplicity, 6
simple, 6
smallest, 99
eigenvector, 5
  principal, 9
elementary symmetric function, 85, 93
epidemic threshold, 38

graph
  bag, 3
  balanced bug, 105
  bipartite, 6, 112
  bug, 3, 104
  color critical, 95
  complement, 4
  complete, 3, 8
  complete bipartite, 3, 8
  complete multipartite, 3, 84, 96
  complete split, 3, 83, 90, 152
  connected, 1
  degenerate, 78
  difference, 49
  domination critical, 136
  double star, 116
  harmonic, 53
  Hoffman-Singleton, 113
  infinite, 25
  infinite bug, 29
  infinite kite, 28, 95, 98
  invariant, 3
  join, 4
  kite, 3, 93, 96
  lollipop, 3, 96
  Moore, 113
  Petersen, 113
  pineapple, 3, 90
  radially maximal, 122, 126
  regular, 1
  rooted product, 19
  semiharmonic, 53
  simple, 1
  spectrally degenerate, 79
  spectrum, 6
  subregular, 64
  surjective split, 129
  threshold, 48
  Turán, 3, 83, 92, 96
  unicyclic, 67, 73, 112
  union, 4
graphon, 151

Printed in the United States
By Bookmasters